Isaiah 58 Mobile Training Institute

Food Resources

of Demonstrating the Love of God to the Nations

All Nations International
www.all-nations.org
isbn: 978-0-9755202-9-1

Isaiah 58 Mobile Training Institute Spiritual Training Manual
© 2016 All Nations International All Rights Reserved

Editors: Teresa Bravata, Ashley Flores, Geraldine Jeffrey, Irene Jensen
Amber Lawton, Karen Offerman, Joe Rodriguez, Melanie Rodriguez,
Virginia Russell, Teresa Skinner

Messages transcribed by: Jennene Jeffrey, Kathy Vanzandt

Artwork: Julian Peter V. Arias, Cheryl Johnson, Joe Rodriguez, Jumi Sabbagh,
Teresa Skinner, George Thomas, Irene Jensen, Adobe Stock, freepik.com, Monique Handall,
Jennene Jeffrey, Virginia Russell, Teresa Skinner, Sylvia Dawn and Annella Whitehead

Unless otherwise indicated, all Scripture quotations are taken from
the Holy Bible, King James Version - Public Domain
Scripture quotations marked (NLV) are taken from the Holy Bible,
New Life version, copyright © Christian Literature International

Cover Art: Julian Peter V. Arias and Eve Lorraine Rivers Trinidad

Isaiah 58 Mobile Training Institute books are available for use in training programs.
For more information or to order additional copies of this manual
email us at allnations@as.net
or contact us at www.all-nations.org
or write to: All Nations International PO Box 26632 Prescott Valley, AZ 86312

Contents

Improved Backyard Chicken Production - cont.

Dedication

We dedicate this manual:

To those who wanted to know... but never had a teacher.

To those who looked for the vision... so that they could run with it.

To those who want to know "What's Next?"

To those who knew they were teachers... but did not know what to teach.

To those who are looking for Christ in Us the Hope of Glory!

May this manual reveal to you Jesus Christ and

May the peace that He has ordained for you be with you always.

The authors.

Acknowledgements

There are so many who are a part of this manual. So many authors and editors transcribers and artists. It has taken more than 40 years to write this manual. Please see individual courses for Acknowledgements

Thank you to those who have:

1 Corinthians 3:6-8 (NLV)

"I planted the seed. Apollos watered it, but it was God Who kept it growing. This shows that the one who plants or the one who waters is not the important one. God is the important One. He makes it grow. The one who plants and the one who waters are alike. Each one will receive his own reward."

Preface

For those of you that are wondering... and are interested in... "What are they doing?

Do you remember Rev. Agnes I. Numer sharing about a school? About a college?

Well....

I was with Agnes in the Philippines in the 90's. She sat with maybe 8 men, all of them leaders of Bible Schools and Leadership ministries in the Philippines. When she shared with them that she was going to build a training center in the Philippines, they all said, "Sister Agnes, we are asking you that you would consider having a mobile training team, so that all of our ministries would benefit from your curriculum." For years we went to different Bible Schools and organizations in the Philippines working with Asian Center for Missions and Tribes and Nations Outreach sharing the principles that God gave to us and through Agnes. Both spiritual and natural teaching like: how to be baptized in the Holy Spirit, how to stay healthy on mission trips, how to raise rabbits and how to flow together and love each other with God's love.

Many of you know, I was so ill for many years. One day, just months before God healed me, He spoke to me and said, "Teresa, you've been doing it wrong. You've been trying to train every person that walked through the door." I was trying to train those who did not want to be trained, those who did not have an ear to hear, a heart to receive, and a heart to obey.

Then God had me read:

2 Timothy 2:2 "And the things that thou hast heard of me among many witnesses, the same commit thou to faithful men, who shall be able to teach others also."

I had cried to the Lord and we had worked with so many and there are so few who took what we learned to the Nations. It seemed unfair to me... how foolish I was. God said, "I trained you... didn't I?" Suddenly I realized how difficult it was to train me... thousands of people later... I definitely repented and was all ears. "Ok, Lord what are you doing?" That was the beginning. The beginning of where we are now.

Shortly before we moved, the Lord told me I would build the school. No, it won't be a school of bricks but a mobile school. No, it won't be for those who will not, it will be for whosoever will... for those who have ears to hear, a heart to receive, a heart to obey. Remember this one?

Habakkuk 2:2 "And the Lord answered me, and said, write the vision, and make it plain upon tables, that he may run that readeth it."

This scripture has resounded inside for many, many years. We started with training paces... but now we must realize it is time. Time to read it and run with it... have you ever thought that in late 7th century BC when the book of Habakkuk was possibly written that there were no IPads or E-readers? This curriculum was made for the modern day "tablet." We have added videos that go right along with the tablet. Now with a couple of wires someone in Africa can use the tablet to have his own Bible School and Training Center.

We are focused on multiplying. We are getting older and time is running away from us. May we spread this truth as far and as fast as God will allow. Empowering Pastors and lay leader the opportunity to teach their people.

What does the curriculum consist of?

Since we have worked with so many Bible Schools our goal will not be to duplicate a regular Bible School with homiletics and hermeneutics... Our goal is:

"What you have heard me say in front of many people, you must teach to faithful men. Then they will be able to teach others also." 2 Timothy 2:2 (NLV)

What God taught us through Agnes and by His Spirit: How to "get the junk out of our lives," how to hear His voice, how to Love the Nations with His Love and how to let Him take us to the Nations.

For others this was "just common stuff." For me it was lifesaving. I would have never made it, I would have never served God and lived through what I have lived through without what God gave through Agnes and also what He gave as a result of her teaching how to hear Him.

I could not have made it: Without the Natural and the Spiritual flowing together... Without Jesus bringing us back to the Father... Without His Be-Attitudes. I would have never been in my right mind without the revelation of Isaiah 26, and how God Himself, made us a new creature, free from the former lords.

As I travel around the world, I see pastors and leaders struggle with what to teach their people. Maybe they have never had Bible School training... and may never be able to afford it.

I appreciate the opportunity to have lived this life changing experience and now, thousands of hours, many authors, editors, artists and volunteers later we are offering this simple Gospel to the world.

We realize this is the first edition. We are still making videos to go along with the teaching. It is simple because Rev. Agnes always ministered the simple, yet profound truth of the Gospel.

Our cry is that God will read this to you... that He will impart His Gospel to your heart and that He will train you, that you will experience the freedom, peace power and ability to demonstrate His Love to the Nations.

May we all work together while there is time.... That He alone may be glorified.

"And this gospel of the kingdom shall be preached in all the world for a witness unto all nations; and then shall the end come." Matthew 24:14

Let Jesus take you to the Nations.....

Teresa Skinner

Project Director

Introduction

All Nations International, a church, and Sommer Haven Ranch International, a humanitarian aid non-profit, are two organizations founded and directed by Rev Agnes I Numer who passed away July 17, 2010 at nearly 95 years of age. She left behind a tremendous legacy after 56 years of ministry. These ministries were birthed out of a revelation God gave her of Isaiah 58. When God showed her this revelation He told her, "This is My plan for My church for the end of time" The Lord showed her planes and trains, warehouses, training and refuge centers, a lot of food distribution and many other things.

It would be difficult to understand the impact this ministry has had over its more than 50 years existence. Almost as difficult as answering, "how many trees are in an apple seed" because that is what this international ministry has done... spread seeds. Many leaders have been given a vision, trained, developed, encouraged and supported. These leaders then have gone and have spawned many ministries around the world. They received a vision, a hope, a plan and principles of God's Kingdom which work and then they passionately put what they had received into practice.

These international, ongoing ministries have learned to know God as Jehovah Jireh. And He provides for them because they are doing His work in His ways. In this training we hope to impart the principles they received and God has so greatly blessed. We give God all the glory. The training is by His Spirit to those who have an ear to hear, a heart to receive and a will to obey.

God showed Rev. Agnes I. Numer a Bible school, a college that would share these principles with the nations. When she visited the Philippines a team of Pastors and leaders asked her to keep the school mobile so that all of their Bible Schools could participate through mobile training team. Isaiah 58 Mobile Training Institute is now available in book and ebook form.

Thank you.

All Nations International

Habakkuk 2:2 (KJV) "And the Lord answered me, and said, Write the vision, and make it plain upon tables, that he may run that readeth it. 3 For the vision is yet for an appointed time, but at the end it shall speak, and not lie: though it tarry, wait for it; because it will surely come, it will not tarry."

2 Timothy 2:2 (KJV) "And the things that thou hast heard of me among many witnesses, the same commit thou to faithful men, who shall be able to teach others also."

CHAPTER 1

Feeding the Hungry

A Practical Guide to the Spiritual Principle of Feeding the Hungry

People are hungry and have unspoken needs. They may be at a time in life that needs God's love. We cannot demonstrate God's Love unless we have His Love. We will not ask for His love unless we have a genuine concern for others. We cannot have a genuine concern if we are selfish and self centered.

Many people want to help, they just don't know where the need is.

We only present the need – God moves on the hearts.

Introduction

Introduction to God's Food Resources

God's kingdom principle of feeding the hungry is not merely an expression of sympathy and charity. The obedience of the command brings the acceptable day of the Lord. God's blessings are given to us and then we give them to the people. As the people receive this blessing from the Lord, they become more and more aware of His love toward them.

We're living in the time of fulfillment of the Word of God. We are living in the last days. God has a special time for us today that He didn't have years ago. We have the privilege of looking into something with the Lord past generations could not enter into. The prophets could not enter into it. No one could enter into it because the Lord said that He kept it for this hours and not before – and no one could say they already knew it. He is getting ready to do something for us to show us how great He is – how mightily He loves all mankind.

So He unfolded this scripture in Isaiah 58.

"Cry aloud, spare not, life up thy voice like a trumpet, and shew my people their transgression, and the house of Jacob their sins. Yet they seek me daily, and delight to know my ways, as a nation that did righteousness, and forsook not the ordinance of their God: they ask of me the ordinances of justice; they take delight in approaching to God.

"Wherefore have we fasted, say they, and thou seest not? Wherefore have we afflicted our soul, and thou takest no knowledge? Behold, in the day of your fast ye find pleasure, and exact all your labours. Behold, ye fast for strife and debate, and to smite with the fist of wickedness: ye shall not fast as ye do this day, to make your voice to be heard on high.

"Is it such a fast that I have chosen? A day for a man to afflict his soul? Is it to bow down his head a bulrush, and to spread sackcloth and ashes under him? Wilt thou call this a fast, and an acceptable day to the Lord?"

This word "acceptable day of the Lord" is very important us to understand. All that is happening all over the world today – the antichrist spirit, all of the evil and the wickedness that is coming – it is important for us to know there is an acceptable day: the day that God will accept a work that He will honor, a work that He will bless.

This isn't Isaiah speaking, it is the Lord speaking. It was a command that the Lord had given. Many years before when the Lord gave me that scripture, "Is it not yet a very little while, and Lebanon shall become a fruitful field, and the fruitful field shall be esteemed as a forest," little did I know it was going to fit in my life.

"Behold, I will do a new thing, now it shall spring forth, shall ye not know it? I will even make a way in the wilderness, and rivers in the desert. The beasts of the field shall honour me, the dragons and the owls: because I give waters in the wilderness, and rivers in the desert, to give drink to my people, my chosen. This people have I formed for myself; they shall shew forth my praise."

Why had God formed us for Himself? To be to Him the praise and the glory. He created us to love Him and to serve Him. As we obey His Word and as we obey His Spirit, we find something else that we probably didn't know. That the love of the Lord, the peace of the Lord, the joy of the Lord is with us. The joy of the Lord is our strength and we have the peace that passeth all understanding from God. For God is bringing His Spirit in this hour in a measure not as it was before, but in the fullness of His power, in the fullness of His Spirit.

This command is very clear – a command of the Lord. We started with a missionary training school. We had five young people and then seven. And we watched the Lord transform our lives and place within us this which He has given of Isaiah 58. Remember it is not man doing it – it is God doing according to His way and if we do it according to His Word, He has promised us that our light shall break forth as the morning, and that our health shall spring forth speedily, and our righteousness, which is Christ our righteousness, will go before us, and His glory will be our rear guard. Now I don't think we lack anything, do you? If God is with us and God is smiling upon us with His life and His presence, He said, "then thou shalt call and I will answer."

What the Lord is wanting us to do is hear His voice, obey His voice, and flow by His Spirit; and if we call, He will do it. He will open up the windows of heaven and pour out on us a blessing we cannot contain. If we will obey His Word, God will do something that we have never heard of before.

And this is what His Word says, "If we give our bread to the hungry, and bring the poor that are cast out to our own house, clothe the naked and hide not thyself from thine own families – help them also."

What I am excited about is when we begin to exercise that which God has given us to do, the Lord is right there with us to bring it to pass. I was very excited with the Lord: before we started our training school all this was hid in my heart, and the Lord didn't let me share it because it wasn't time to share it. But when He started the training school

He began to pen it before us, as to what He required of us. He gave me many things that we are seeing fulfilled right now.

Today God is doing mighty things because of the circumstances in the world, because of the famine in the world. We are seeing all this coming upon us, and we realize that God has a plan. And because we know He has a plan, we are going to listen to him and let Him help us with that plan. It's not our plan, it's God's plan. And as we hear what He is saying, He puts in our heart to help others.

Something begins to happen to us when we begin to obey the Word of God. We begin to realize that it's time that others should know how great our God is. That what He said He will do, He will do. We need to just open our ears and hear it, and obey it and receive it. In our training school the things that we teach are to have an ear to hear, a heart to receive, and heart to obey. And to learn to flow with the leading the Spirit of the Lord. And when you do, something happens.

One of our young men was walking down the road and the Lord just spoke to his heart, and there was a man coming on a bicycle with his little dog beside him, and he stopped to talk to him. And he said that there were families up the street from us who were hungry, and they didn't have any food, and they didn't have any money, and the dogs were hungry. And so they came to the house. Our young man said, "I don't know how you feel about his, but I think we need to take food to these people."

So that was the beginning of the food ministry. And that same young man was going down a country road looking for hay for our animals and woman and her son were walking down this road. And the Lord said, "Stop and talk to them." And so he stopped and began to talk. The mother couldn't speak English but the little boy could. Her husband has just had heart surgery and was disabled and very, very ill, and they didn't have any food. So we fixed them some food. The Lord delivered them from alcoholism and this family became the greatest distributors of food – 350 families in this valley. It began with a little food box.

Sometimes we don't intend to do something, but the Lord intends for us to do it. There comes a need and the Lord drops it in our heart, and all at once we realize that God is doing something for us. The daughter of that first family came to live with us and God transformed her life. He changed the life of her parents and gave her something to live for. If he had not stopped as the Spirit of the Lord had directed him, we wouldn't have her with us now.

Sometimes it's just a little nudge – just a little thing – and we want to shove it aside and say, well, it's not important. We don't realize how important it is until we pay attention to it. Then we begin to see it expand and expand, and God begins to enlarge it, and it becomes a great thing that He would have us to do.

I believe it is very urgent for us to receive a burden to feed the hungry. Even in this city, in all of Los Angeles, there's a lot of people on the streets. Not of their own fault – no work, no house. They can't afford the high rent and the housing. They can't afford to eat. We help many families. If they pay the rent, they don't have money to buy the food. We help support them in food. They wouldn't eat if we didn't help them. But we go by the leading of the Lord. He directs us who to help and who not to help. You see, there are people you want to help and you can go on helping and helping and they are never changed. But God wants to tune us with Him, and He will direct us whether to help them or not to help them.

Something happens with us. The Lord made us very aware it was Him doing it, and as we obey Him, He will receive the glory. If God doesn't receive the glory, there's not anything accomplished. If man receives the glory, God is not pleased with it. God wants to receive the glory Himself. If we let Him direct us in helping others, if we are careful in that we give Him the praise and we thank Him and something starts happening. But even as we feel inside of us the changes that are taking place, He's also changing the hearts of the people. One thing that we tell people: God has provided it and give Him glory and thank Him for it and praise Him for it. It turns their attention to the Lord. It turns their attention that God is a great God and He does love them. And pretty soon they open their hearts and they open their minds – they hear what the Lord is saying and their lives are changed.

We have tremendous testimonies of how God has changed lives by this very principle and the very fact that we obey the Word of God and do what God requires of us.

And the most amazing part of it is God says He is with us. He is with us! He will lead us to those people who have needs… not only hungry, but spiritually hungry – both problems. Burdens and heartaches and sorrows. God can use us to speak to them and pray for them and see their lives changed.

So the Lord has given us this plan that we might hear His voice and obey Him and let Him prepare us. We stress so much being led by the Spirit of God, not by our flesh, but by the Spirit of the Lord. The Lord leads us more than we think He does. The little still, small voice within us.

A lot of times we think it is an impression: but it really isn't just a little impression. It's God directing us to do something. It may be strange to us. It may be different than anything we've ever done. But when we do it we see the hand of the Lord moving in a very special way.

We pray over the food as it comes in, we pray over that food as we give it out, and we ask not only for the Lord to bless the food, but to bless the people who receive the food, and as the blessing of the Lord goes out with the food and the people are blessed by it, the Lord has an open door to minister to them.

We've many wonderful testimonies of the changing of lives because we give food to the hungry. Finally the Lord gave me a very mighty message about a cup of cold water in His name. How much we need to just be tuned and sensitive to these things because as things get worse – and they will get worse – God will have his people who are there to help those in need. Not only to help those in this country and in our valley, but we help people in other countries. God gave us the plan, and I know it is a very mighty plan, because God gave it and it works by His Spirit.

In the villages of these countries where there is poverty, the people are very, very illiterate with no provision, not even knowing how to grow crops or any such thing. And God showed us this. We took seeds – garden seeds – and taught them how to grow their crops, then they would have food. Then we taught them health and hygiene. And the simple things that you and I know so well – they didn't know it! We went over into Nigeria and took a team with us and they said to us, "Nigerians don't give anything... they don't give anything! I don't think they have anything to give." We were in a chateau and here comes some of the people with the food. "We want you to have this food." One of our young men when they said this to him, said, "But God."

When we took six containers of what into the Philippines, the Filipino people said, "Filipinos don't eat wheat." Some of the people said, "Well, you won't see the miracles in the Philippines that you saw in America because the Philippines are different." But you see, God isn't different. And God did the same things for them that God did for us in America. So God just opened the doors for us. All of it was taken through without cost. Everybody donated everything – six containers of wheat donated to the Philippines!

We had to wait almost a year before we approached them because we knew God had a time. Let me tell you, during that year, there were two laws that were passed!

The Lord kept saying to me, "There's a law. There's a law..." Now, I didn't know where that law was or what He meant, but inside of me was, "There's a law."

Right in the harbor of Los Angeles was our wheat being stored free of charge. And two months before we put the wheat in storage, they passed a law that they can take donated food to third world countries without charge.

And another law was passed in the Philippines. One of the governors met with a man that the Lord appointed to take that wheat into the Philippines, and he said there was a law just passed and the law gave the governors of the provinces of the Philippines the right to receive, free of charge, donated food for the Philippines. God opened that door through that wheat to clear the way for other things to be brought into the Philippines. God is great!

You know, we look at a little tiny thing and we think it's small. But it's not small if God is in it. Little is much if God is there. Much little if He's not there. So you might think, "Well, what am I going to do? I don't have the money." We didn't have any money! We just obeyed the Lord and the Lord began to expand. What we're saying to you is: let the Lord direct you in whatever He wants you to do.

Now many people come to us, "We don't know how to get started. We don't know how to begin. We don't know what to do." But there's always that beginning that God gives to us. Maybe it's very small and we don't feel it's anything. But all at once we watch God expand it and we become aware that it's God's provision for ourselves and for others.

The provision that God has given to us is now so great. We started with two orders of food and three quarters of it was rotten when we started. But we sorted and worked and learned how to choose the good from the bad and throw the bad away. God trained us Himself with the food.

Today we don't have that problem. We have lettuce that needs the outer leaves taken off or something like that, but we learned the value of food. You learn to appreciate it because God is giving it to you. It's His blessing to you and His blessing for you to give to others. You know, once you begin it, you'll get so blessed and so excited that you'll wonder why you waited so long to do it!! Because the blessing of the Lord will overtake you and inside you are going to feel something that you didn't think was possible at all, but God is saying to you, "Test Me and try Me, prove Me, and I will show you what I will do."

God is saying today, "Is it not to deal thy bread to the hungry?" The very first thing we gave to the people was bread. The first provision to us wasn't an accident, because bread is the staff of life.

So if God puts on your heart to help others, I would say thank Him, give Him the praise and the glory, and as you give it and share it with others, be careful that they know it comes from the Lord – that it is His provision. God uses people – yes, He does! He doesn't rain manna from heaven, but God moves on the hearts of the people that don't know Him to give of their abundance and what happens? God opens up the windows of heaven and blesses them!

We've had businesses that have had a hard time and they started giving to us and God began to bless them and they gave us fresh things because they didn't have any leftovers! We rejoice in the fact that the Lord has given them all these things because they had obeyed the Lord. Many of them were about to fold up and God began to bless them and multiplied unto them the things He had for them. You know we find many, many people that God has so changed their lives that it is just amazing. There's countless numbers of people that God changed their lives through the fact that they gave us food from their markets that they were going to throw away. And it's good food – food from the King's table in abundance for you, and God has done it.

Just take that step… if it's a little bit, just take it. Let the Lord be honored and glorified, and He'll do something and you'll wonder how He's going to do it! The Lord will show you how. He didn't just save us to sit the in the pews. He saved us that we might give this Gospel of the Kingdom to everyone everywhere that He might be glorified. I hope that God has imparted into your heart, into the innermost part of your being, that anointing that He can use you, that you might be a vessel of honor unto Him. Don't get discouraged about it. It will come very fast when you do it because God is in charge, not man. If God is in charge, you can expect miracles! You know, I will grant you one thing, when you start it, you will get so excited about it that you won't be able to contain it!

I can tell you that once we flow by the Spirit of God, our bodies begin to feel the unity and the flow of the Spirit of God. Not exhaustion, but uplifting by the Spirit of God and God will change our lives quickly as we just do these things that He has commanded us to do.

God wants to call upon us to be strong and help others because the need is going to increase. And God will call upon us. It's the Christians God wants to feed the hungry – not the government. It's the church that God called to feed the hungry, not our government. And God will bless us and give it to us and multiply it unto us as we give it out until we will be very, very astounded at the provision of the Lord. Everything that He does through us He will bless if we give Him the glory for what He has done for us. This is the way the Lord used us in the beginning and now the multiplication is beyond anything that we can comprehend!

You'll really get excited because when God does it, you're going to be happy, and I know from my own experience how mighty it is to obey the Lord, and let Him move through you to accomplish His purpose. I praise God! There's so much… yes, there's just so much… you just need to start and do it!

Rev. Agnes Numer
Founder and Director
Sommer Haven International Ministries

Who May Want to Help?

Who May Want to Help?

Your Church

Village, Community, Family, Friends, Farmers and Ranchers, Owners and Managers, Grocery Stores, Cold Storage Facilities, Rice, Flour, Grain Companies, Large Corporations, Small Businesses.

Many people want to give, they just don't know where the need is. We only present the need – God moves on the hearts.

Joyful Offering

A joyful offering is a set aside by your church to collect food that will be given to those in need.

It is joyful because everyone can give.

If you have an abundance, give freely. If you only have two of a certain thing give one, chickens, bowls of rice, bags of flour, ears of corn, cans of milk.

Ranchers and Farmers

Sometimes farmers and ranchers have surplus food. This may include livestock, poultry, eggs, grains, dairy, and produce.

This food may be non-marketable culls, an over abundance of crops or after harvest gleaning.

The farmer may want to pick these crops himself, or he may allow your group to come into his field and pick.

It is important to develop good relationships with those who give to you. Be careful not to destroy the plants or the fields.

Community Gardens

Sometimes a village or a **community garden** are the only possible means of collecting food for a food ministry. People join together to plant a garden – the abundance is distributed to the needy. Everyone from the youngest to the oldest, richest to poorest can be involved. Lend a hand, water plants, weed the garden, collect and distribute food. Community gardens also provide training for families who to produce their own food. Gardens can be used to supplement any food ministry.

Many people want to give, they just want to know where the need is.

We **only** present the need – **God moves on the hearts.**

Presenting the Need

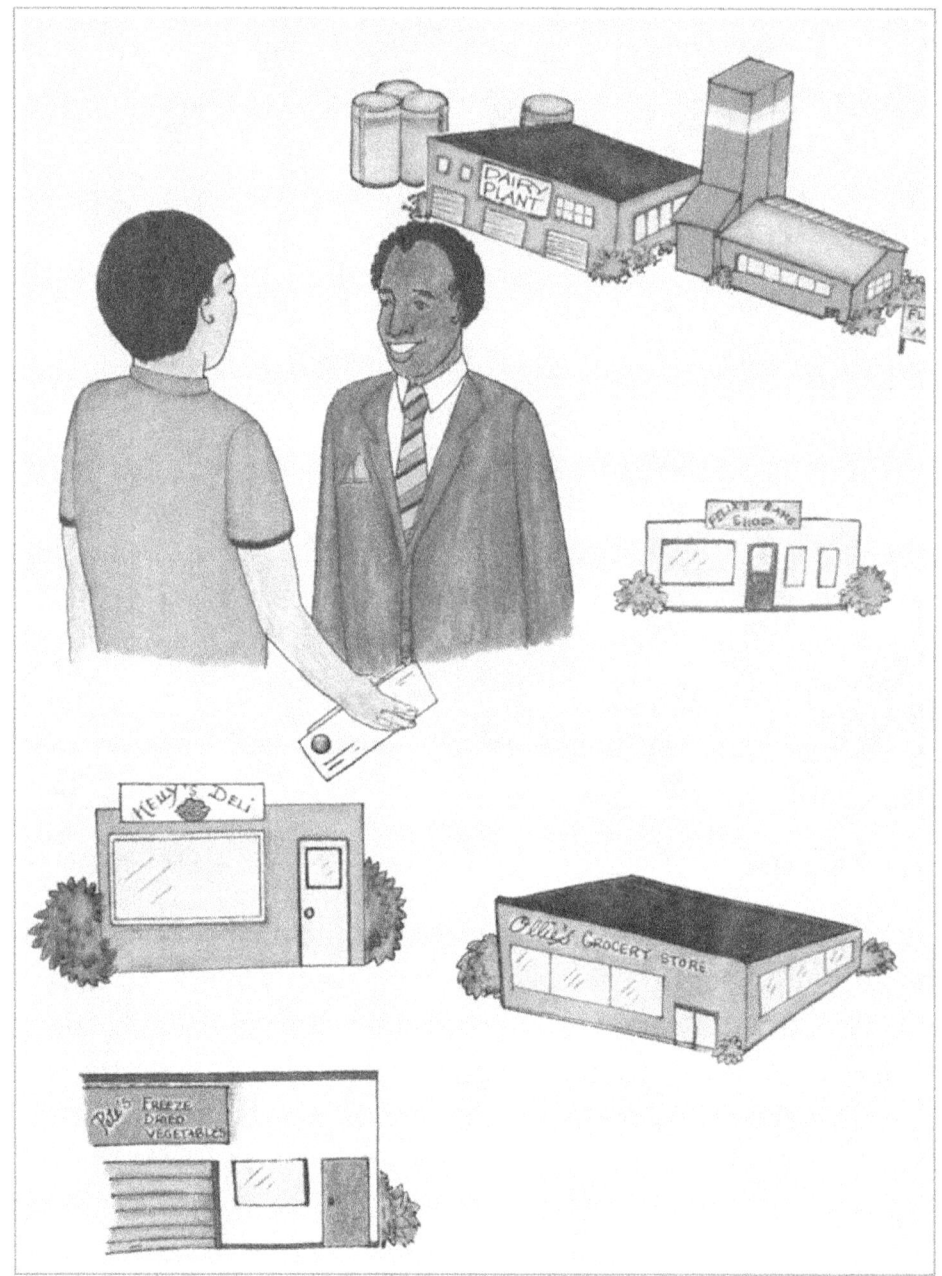

Visit local stores and businesses, and talk to the owners or managers. Write to the owners of store chains, **expressing the needs of your community and how their store may want to help.** Assure them that absolutely nothing will be sold, traded, or wasted.

Go with the anointing, knowing that God is sending you!

It is important to:

Carry a letterhead showing you are helping those who have a real need.

Be friendly, smile. The joy of the Lord is with you.

When possible, take someone with you. Let your words uplift and encourage one another. Pray together for guidance and pray for the people that you will come in contact with that day.

It is important not to:

Give a careless presentation or use slang language.

Wear untidy, torn or dirty clothes.

Do not become discouraged when someone says no!

Many people desire to give, and but may not know where the needs are.

Grocery stores, cold storage facilities, rice, flour, and grain companies, large corporations and small business. **We present the need – God moves on the hearts.**

Delivery drivers may also be helpful.

When you speak with them encourage
them to share with their managers what you are doing.

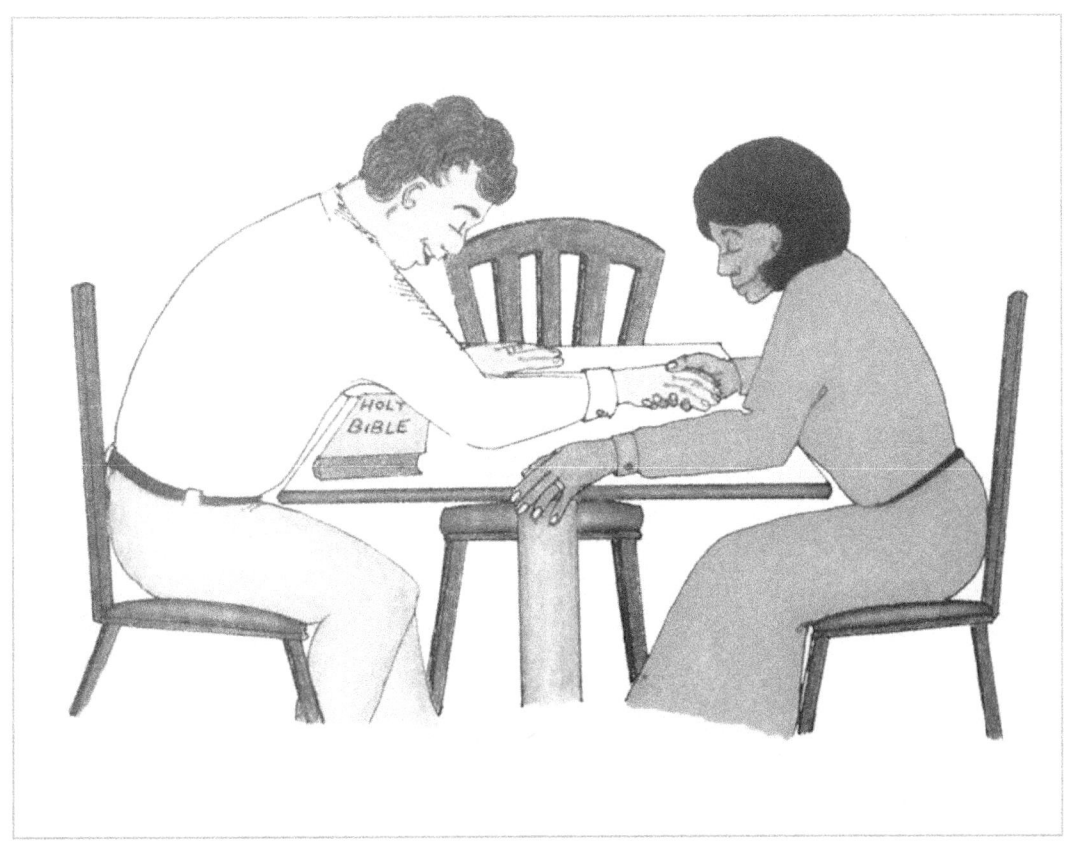

Ask the Lord for His direction daily. You don't have to wait to go overseas to be a missionary. God can use you to lend a hand, pray for someone, say a kind word, **showing others that Jesus can live in them, too!**

When the Truck Comes In

Once the daily route is established, don't hesitate to follow the Lord's leading. Maybe, God will send you at a different time to one of your regular stops, or to a totally new store.

With God the Possibilities are Unlimited!

In the Morning Before You Start:

Gather boxes you will need for the day.

Check the oil in your vehicle.

Check the gas.

Write down the mileage.

Often food will be collected from the back of the store. Talk to the clerks who work in the back so you will know exactly where and when to collect the food. It is important not to be late for any appointment.

Be courteous. Make sure your appearance is clean and neat. Always clean up spills that you make and perhaps those that you didn't make.

To save time later on, bring home as little garbage as possible. Also sort out anything that might damage the produce.

Show by your words and actions that everything you do is because of the love of God in your heart for others. Remain grateful to God for all of His provisions. Never allow complaining, but keep good attitudes about God's blessings.

Do you have any extra time between stops? Think: Is there something I can do?

Minor maintenance on the car. Sort produce. Run an errand in town.

Once you get home the job just begins!

Show inventory sheet and all items received to those who have the responsibility over the food ministry. They will know the needs of the home training center and those who will receive the food. This will help them decide where to store the food and where it will be distributed.

Write down the mileage.

Return all receipts to office personnel.

Weigh all received items.

Report any vehicle malfunctions. Report social concerns, prayer requests, or special needs of those you are in contact with during the day. Give "praise reports," others will also want to give God glory for his provisions that day. Pray for God's daily provisions.

You may receive items which may be "Just what was needed". A special treat for someone's birthday, anniversary, or special need. Special care should be taken that these items be shown to those in charge of the food ministry so that it will be given to the right person.

If you have livestock, keep in mind that they might enjoy that box of lettuce leaves or other produce that is not humanly consumable, but is palatable for animals.

Handling and Storing Food - 1

Wash your hands with soap and water before handling food!
When handling food keep your hands away from your hair and mouth using disposable gloves when necessary.

Proper hand-washing:

Wet your hands

Use soap to lather, rub between your fingers and scrub your nails for 20 seconds.

Rinse your hands

Towel or air dry turn off the tap with your sleeve or towel.

A Job Worth Doing is Worth Doing Right!

Leaving your job for someone else to do can be discouraging. Clean up spills and remove empty boxes. Return crates and extra containers to the right places. Let's make the effort to flow together to meet the needs of others.

We must realize how important it is not to complain, but to have the right attitude in all we do. The entire provision relies on God's blessings, we must be grateful.

31

When the food is re-boxed, pick up remaining food so that nothing is wasted.

32

Organize food so that the oldest items are in the front and easiest to reach. Mark bread, pastry and potato chip boxes with the correct code date to aid proper distribution while food is fresh.

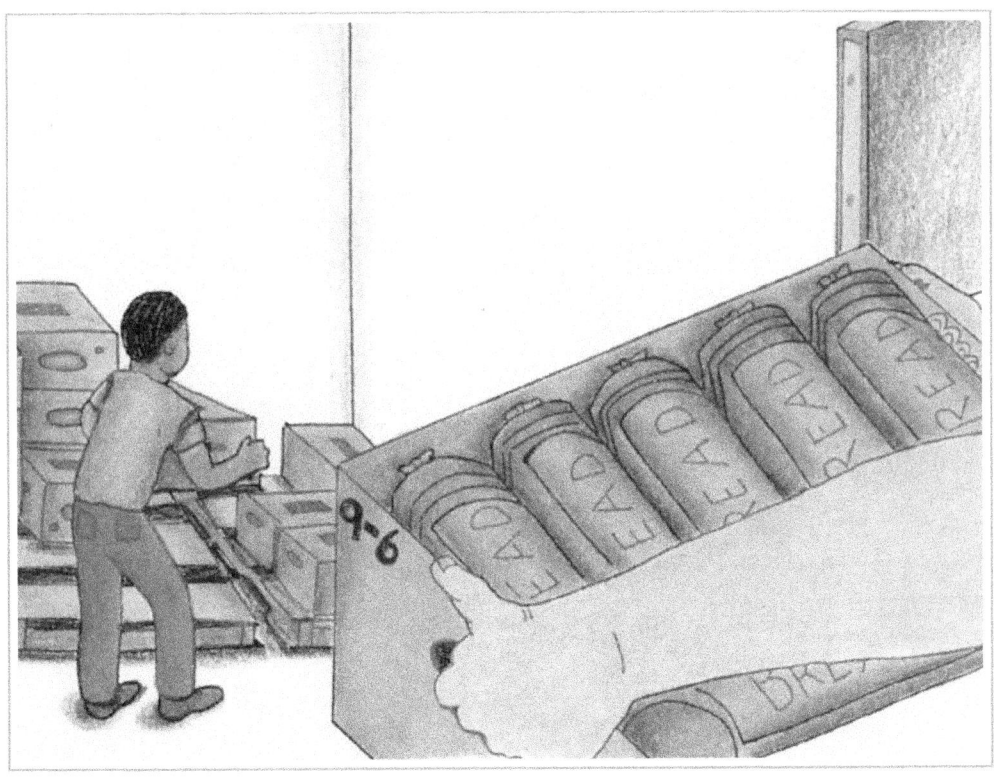

As a safety measure, when lifting boxes, use your knees to lift, rather than your back.

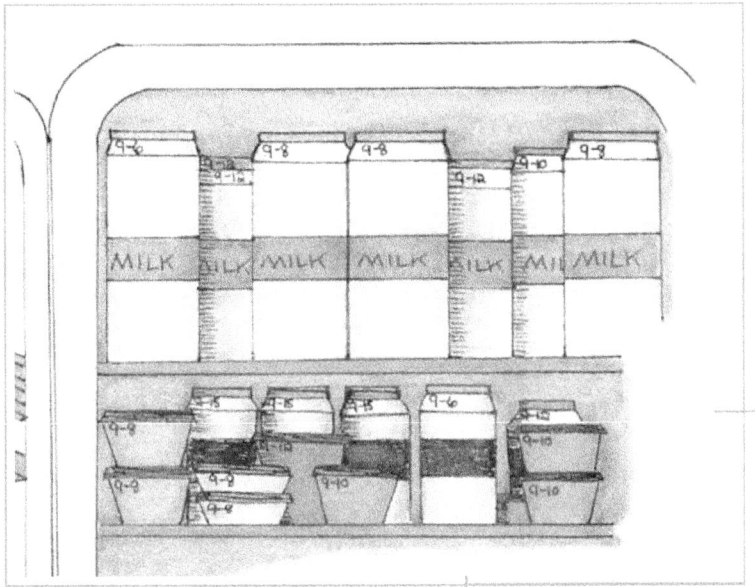

Transfer food items such as milk from a cracked or leaky jug, into fresh, clean, sealable containers marked with the correct date.

Separate all dairy products always stacking the oldest items towards the front.

If you bring home eggs, sort out the cracked ones. See if any need to be wiped with a clean cloth before putting them away. Rotten egg and sour diary odors make a clean refrigerator smell unpleasant.

Food absorbs odors. Keep refrigerator units smelling fresh, the way you would like your food to taste.

Whatever spills you make, clean them up right away. Leave the area where you were working cleaner than when you got there.

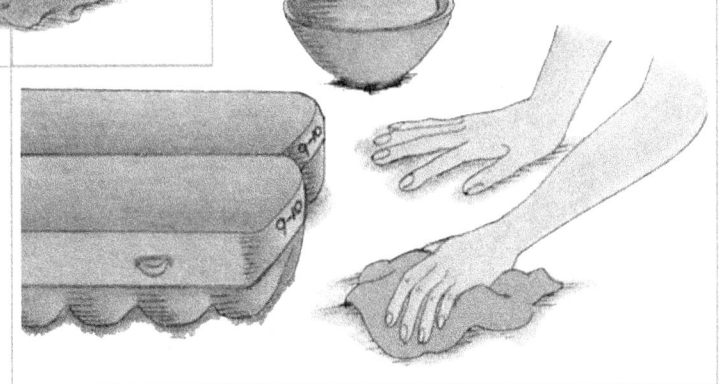

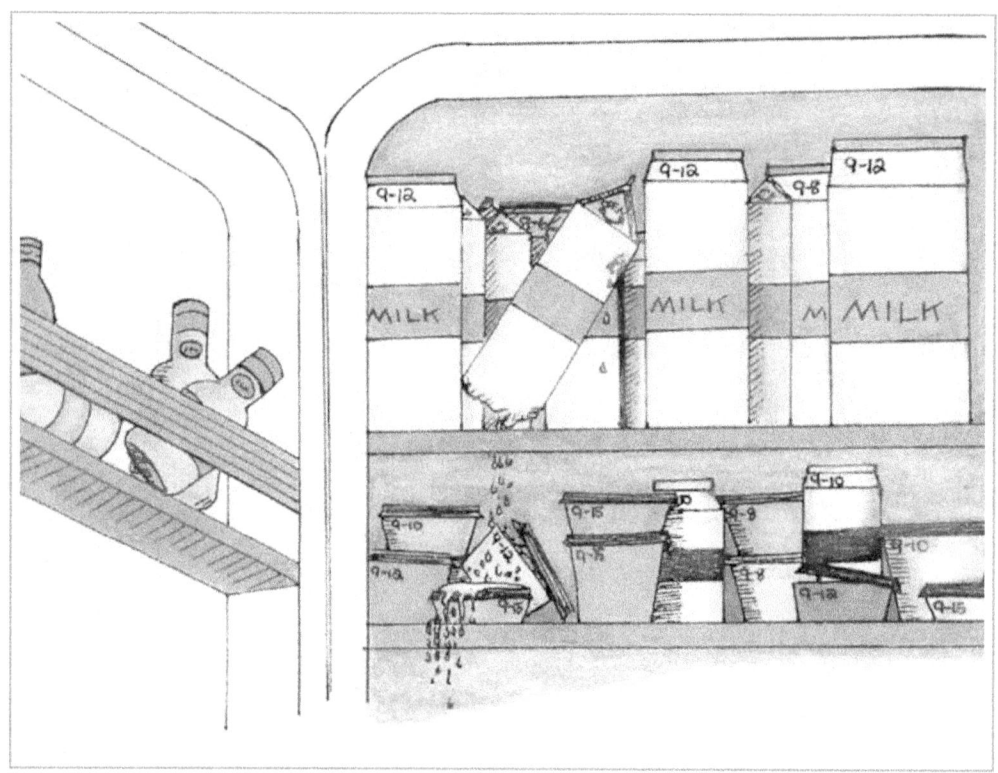

How you put food items away will either bless or discourage others. Train those who give food and those who receive the food, to separate damaged food items. This will keep the good food from spoiling.

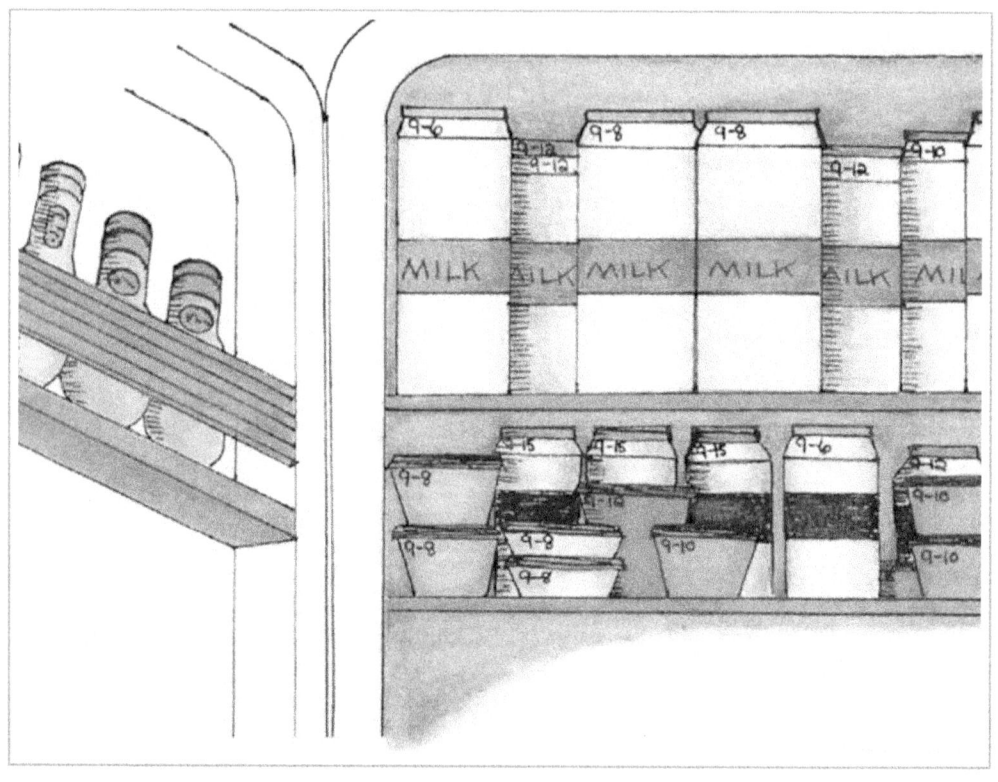

This food must be "worked up" and cooked right away. For example: Bad spots are removed from peaches and what is left is used for peach pie.

Let's make the effort to flow together to meet the needs of others.

Handling and Storing Food - 2

Wash your hands with soap and water before handling food! Keep hands away from your hair and mouth using disposable gloves when necessary.

Proper hand-washing:
Wet hands

Use soap to lather, rub between your fingers and scrub your nails for 20 seconds.

Rinse your hands

Towel or air dry turn off the tap with your sleeve or towel.

Always wash your hands with soap and water before handling food!

All food with strong odors, such as onions and broccoli, should be kept separate. This way they will not affect the flavor of other fruits and vegetables.

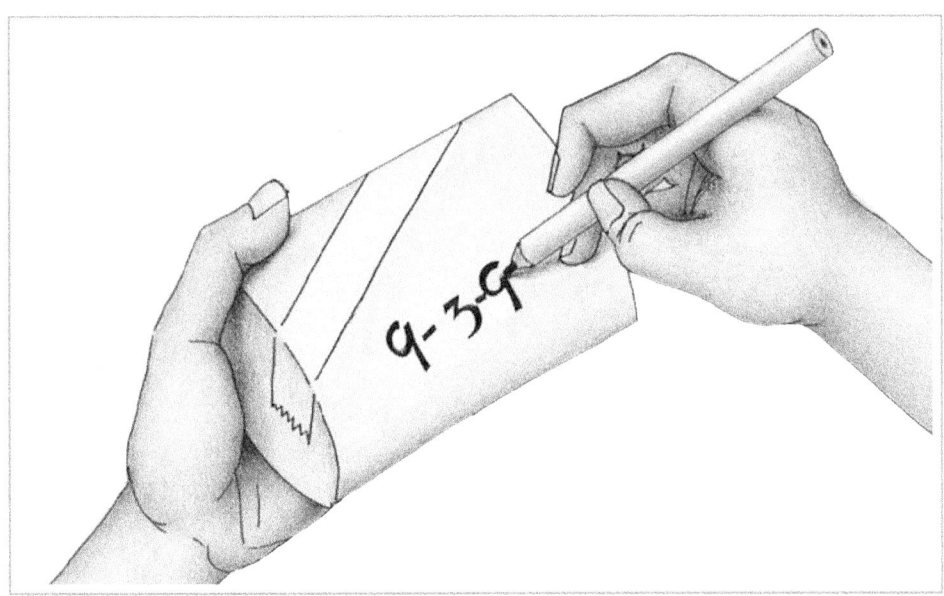

Mark code dates on all incoming items to ensure that they will be given out while they are still fresh.

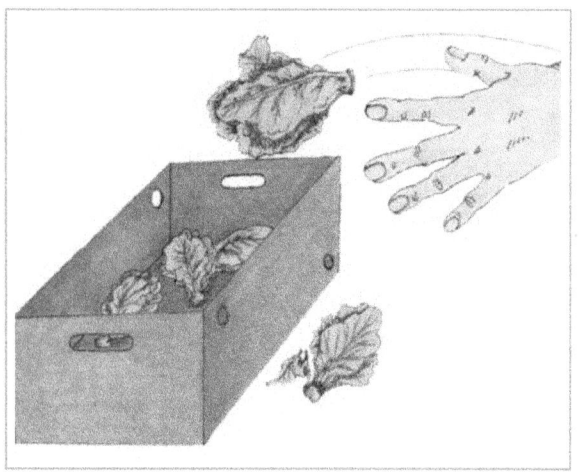

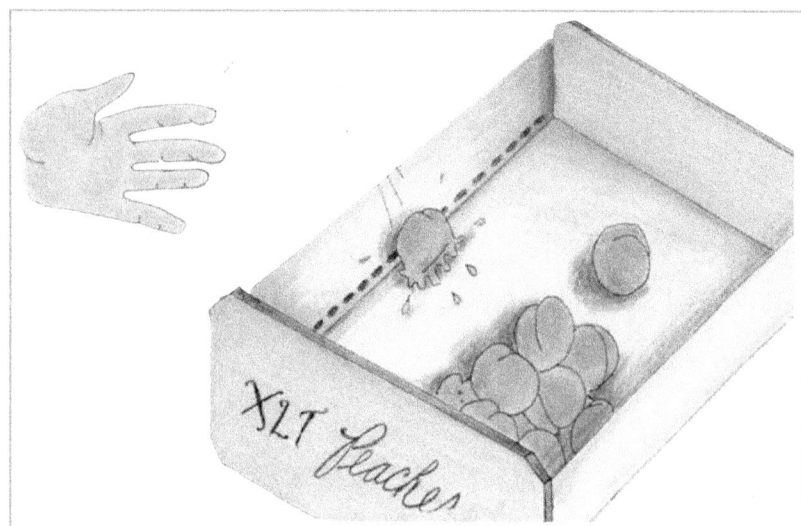

If food is thrown or roughly placed into boxes it will become damaged.

The results may not be seen immediately, but, by the time the recipient receives the food, brown spots will be apparent and the food will spoil quickly.

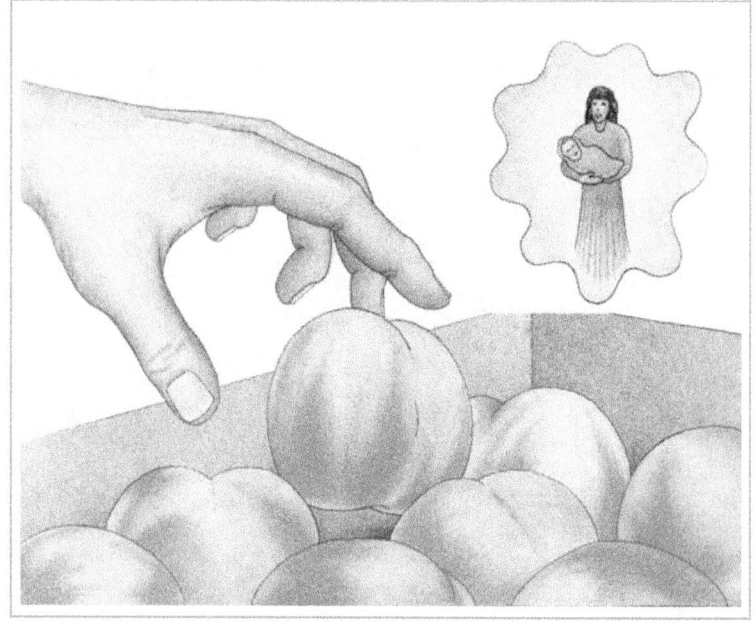

The level of the fruit should not exceed the height of the box. Otherwise the food will be smashed when another box is set on top of it.

It is better to store fragile fruit with a lid. This will help when stacking.

 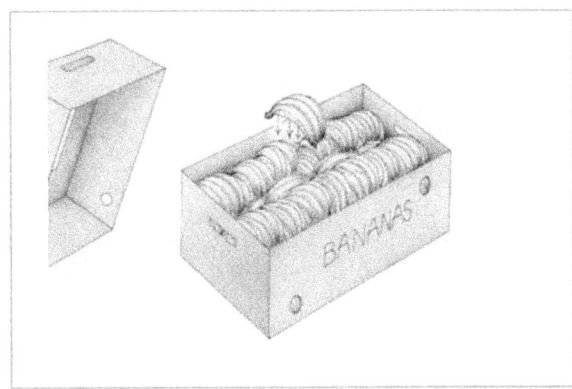

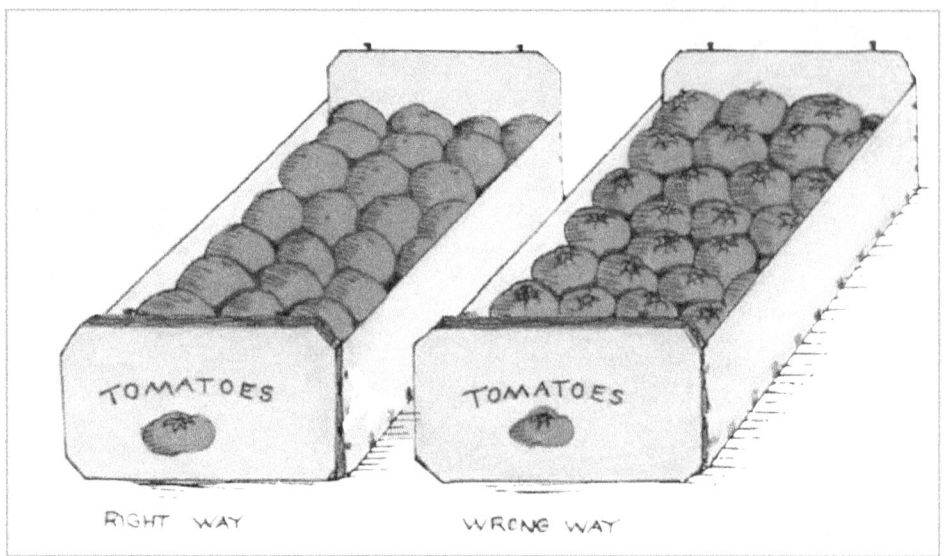

Store tomatoes with stems down. If tomatoes are stored with stems up moisture collects at the stem and causes mold to grow quickly and the stems will puncture the tomatoes stored on top of them. Take care so that they will not bruise and spoil quickly.

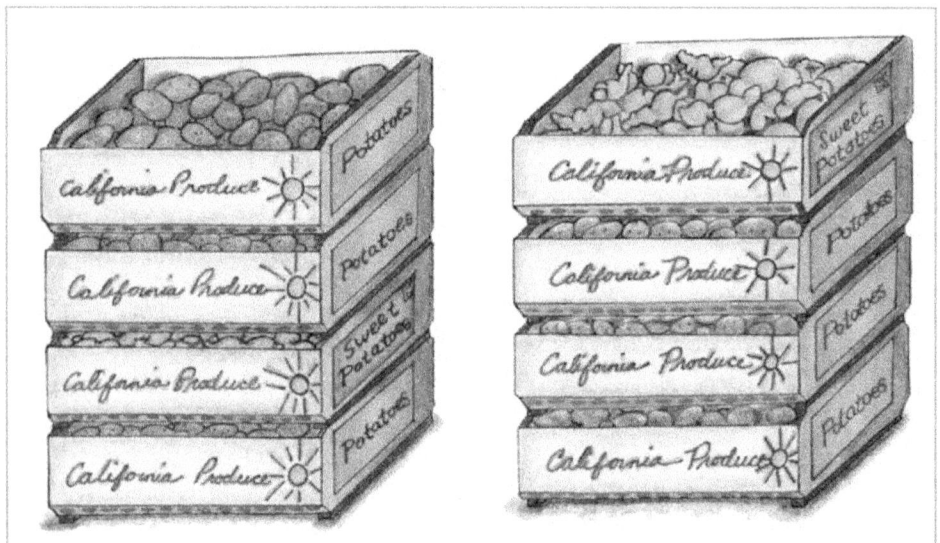

Sweet potatoes need extra air circulation. Place boxes so they are not neglected or overlooked during storage.

Place food with strong odors, in a separate bag so they do not affect the flavor of other fruits and vegetables.

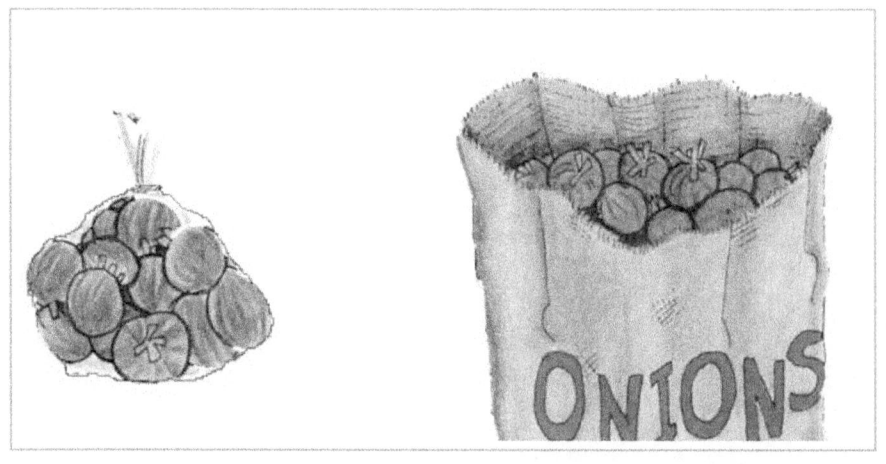

Stack boxes neatly. Be careful not to put them in walkways. Put the heavier boxes on the bottom, and the lighter boxes on top of them. Interlock each stack so that the boxes are less likely to fall. Sometimes there is a diagram that shows how to interlock, stamped on top of the box.

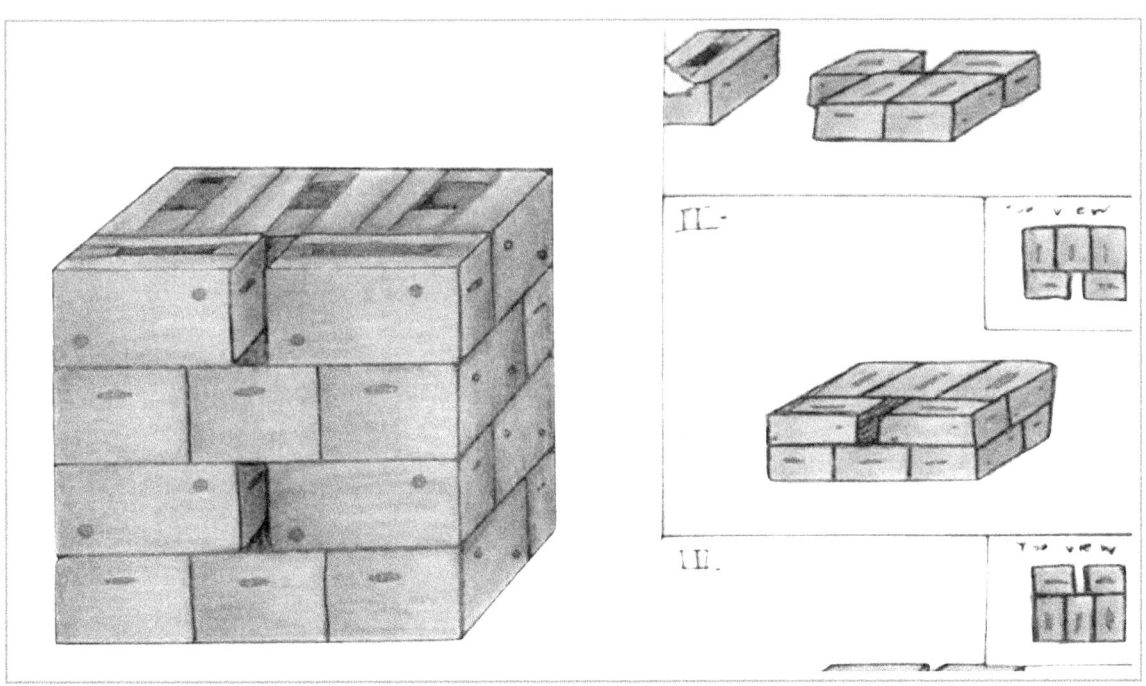

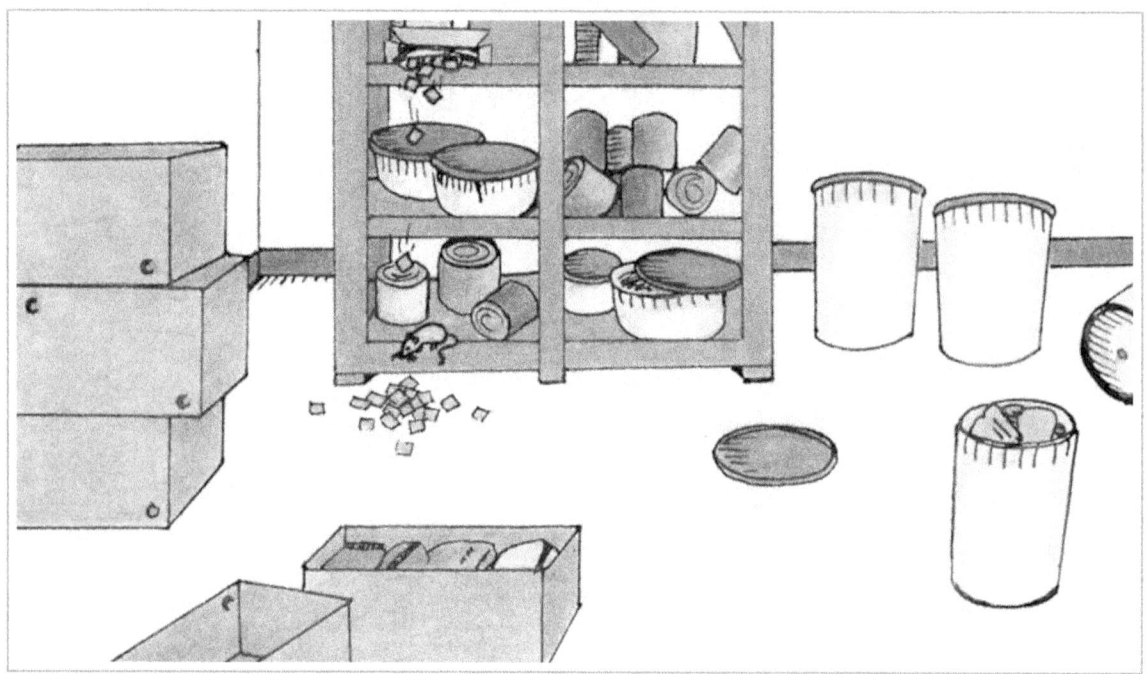

Canned items and dry good should be stored in a cool dry place, off the ground, away from mice, and in air tight containers. These food items are especially useful for needy families, and for families that must be sustained for long periods of time.

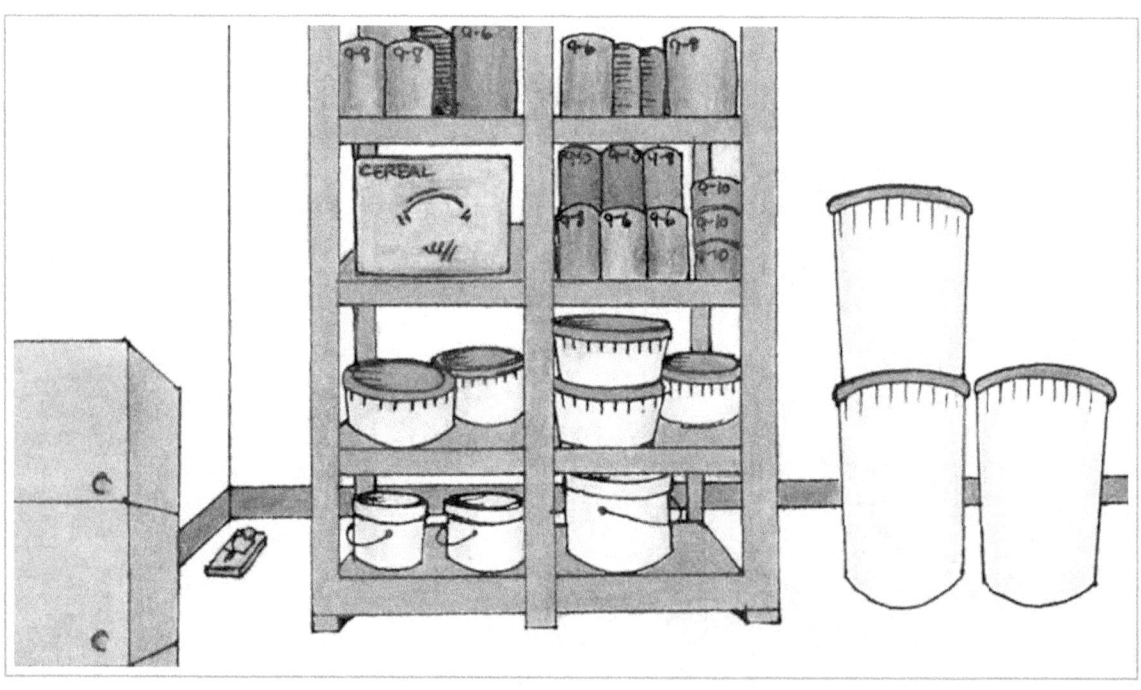

Personal Food Orders

How do we give God's food?

Start with a clean box. Place a clean newspaper or a liner in the bottom of the box. Keep food areas clean away from dirt, germs, and flies. Do not put food on the ground and always observe health code laws

As you make the order, ask the Lord to bless the food. **Pray for the people who will receive the food.** Remember:the food is a blessing from the Lord.

1. How much food do we have?

2. How many people will we feed?

3. How long will this food last?

Sometimes you will only have enough food to give the same amount to each family.

If You Have More Food:

1.   Here are some points to consider when deciding how much food to give to each family.

2.   How many people will this food order actually need to feed?

3.   Babies do not eat much and they only eat limited items.

4.   Teenagers, especially boys, generally eat a lot.

5.   Relatives or friends may also live with the family.

6.   Is this a ministry house that will feed a lot of people?

7.   Is someone in the family on a special diet that requires specific food items? You need to give a greater amount of those items. Is there food coming to this family from any other source? Or will this food be all that the family will have to eat?

8. Does the family have a way to cook the food?

9. Does the family have a refrigerator and electricity?

10. How many days will this food order need to last?

11. Does the family know how to prepare the food that you are going to give to them?

12. Will they be willing to try the different food items they receive?

13. Fresh fresh fruits and vegetables are important in everyone's diet!

Use the most perishable items first. Each family will need plenty of time to handle the food before it spoils.

Train those who give food and also those who receive food, to separate damaged food items. This will help keep the good food from spoiling. This food must be "worked up" and cooked right away. For example: Bad spots are removed from peaches and what is left is used to make peach pie.

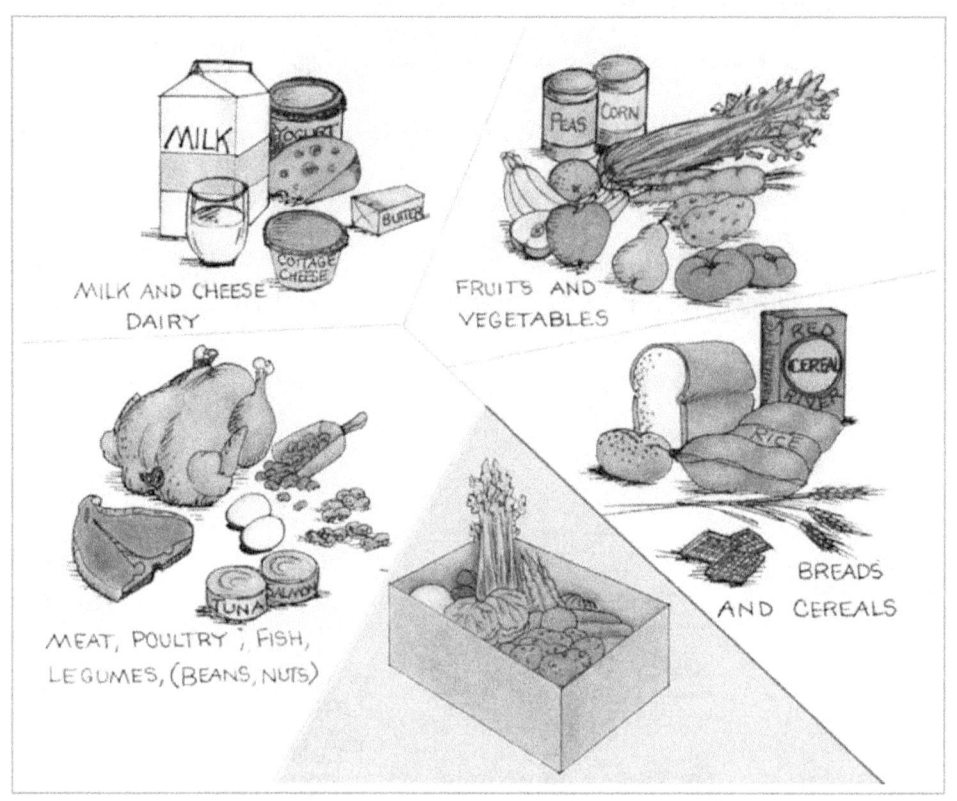

Give balanced meals from the five food groups.

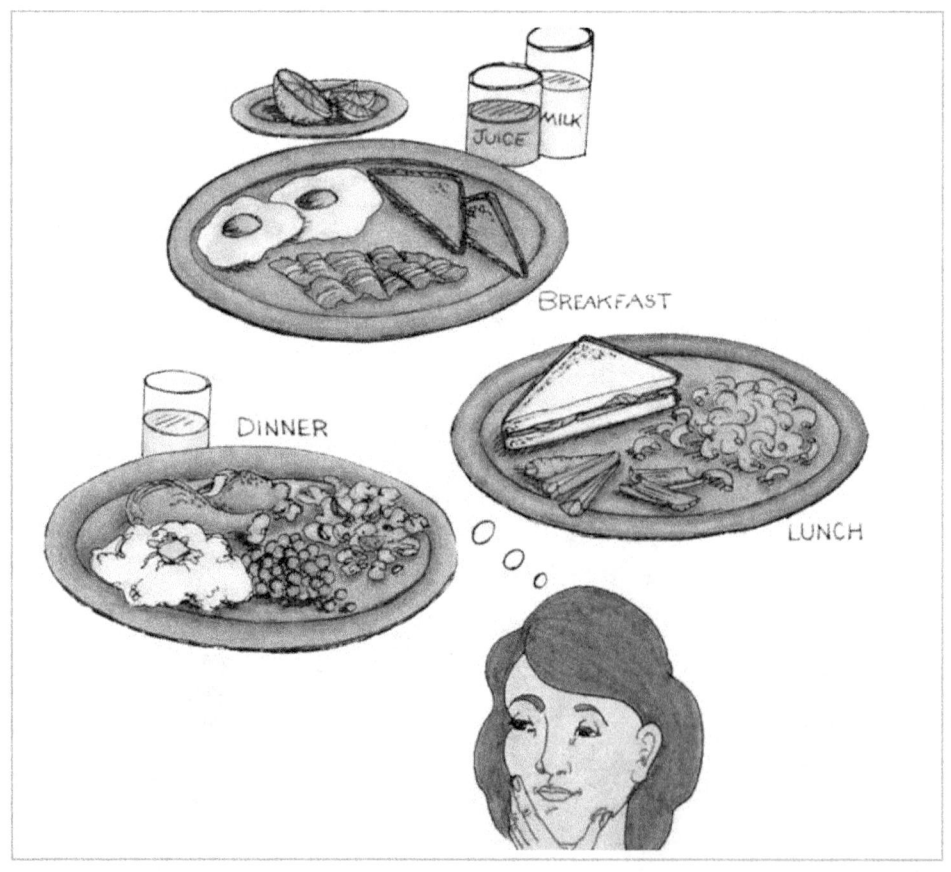

48

All food with strong odors such as onions and broccoli should be placed separately in a bag. This way they will not affect the tastes of other fruits and vegetables.

Separate produce, food items that need to be refrigerated, dairy products, bread and potato chips, into different boxes or bags. This will help to prevent the food from being overlooked or becoming damaged.

If you are putting a food order in the trunk of someone's car, always check to make sure there is nothing that might contaminate the food. Stored fuel, detergent and bleach are all items that, if mixed with food, will make it harmful to eat!

Being friendly and courteous will help each family know that you genuinely care about their needs. Let them know that the reason why you care, you care because God cares. We are not drawing people to ourselves, we are drawing people to Jesus Christ – by His love.

Take time to ask questions:

Does the family know how to prepare the different kinds of food that you give them?

Suggest recipes to each family for foods that are new to them.

Encourage them to bring other families that may have needs.

Show the family:

Which food will need refrigeration.

How long each item is expected to last.

How to prevent unnecessary spoilage.

What signs to look for that will show you an item needs to be used quickly.

Which foods are necessary to use right away.

How to put together nutritious meals.

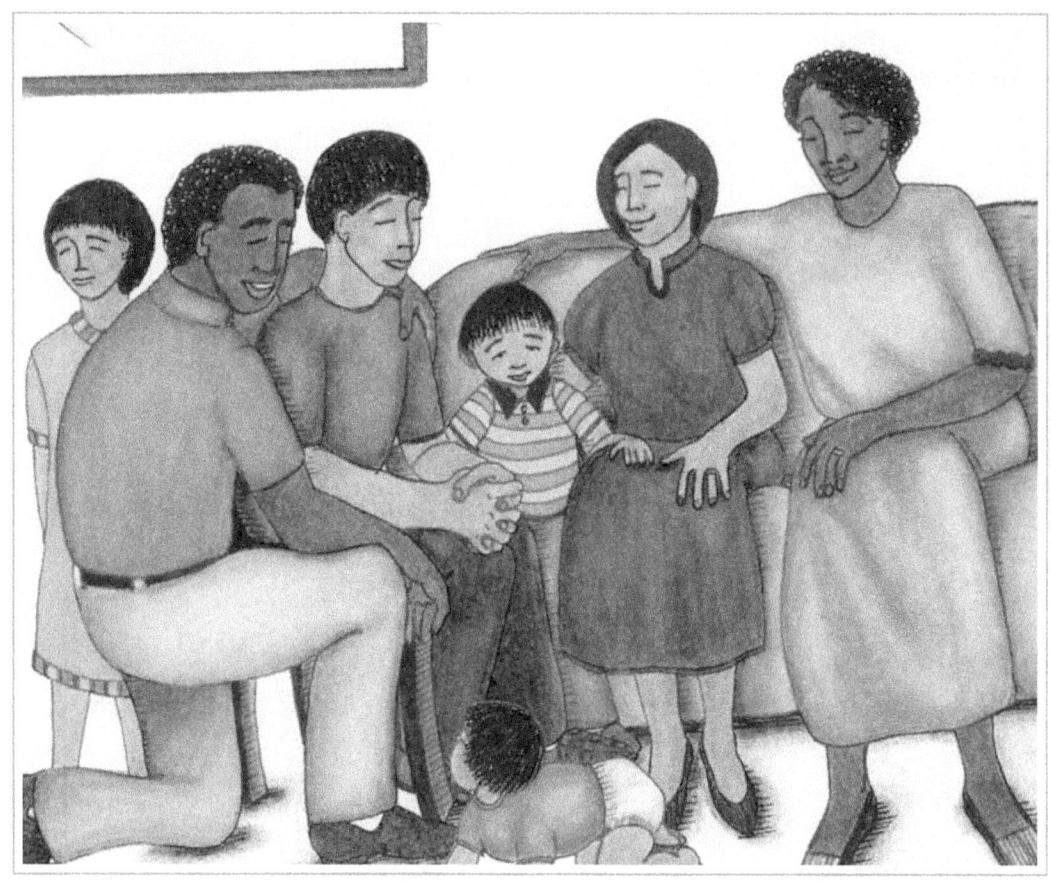

Perhaps someone in the family has lost their job, or there has been a fire, or some other kind of tragedy. These are special times when people need for us to be sensitive and compassionate. They may need clothing, cooking utensils or furniture.

These are families that may have great needs. Give the extra canned and dry goods, variety meats, cheeses – staple food items. If they have a freezer, you can give them frozen foods. Be prepared by storing extra items for these situations.

As you give, pray for God's blessings on the food. We are stewards of His blessings. As we give with His love, He is faithful to move within the families who will receive the food.

You may feel to pray or share a word from the Lord before giving the food. Remember, some people and families want and need prayer.

God's gifts are an expression of His love for man. As we express His love, other will feel His love for them, and know they can come to Him!

Acknowledgments

Isaiah 58 Mobile Training Institute Spiritual Training Manual
© 2016 All Nations International All Rights Reserved
Used by Permission

Author: Teresa Skinner

Artists:

Third Edition: George Thomas, Cheryl Johnson,

First Edition: Edith Batagas, Wayne Sigler

Special Thanks:

Rev. Agnes I Numer, George Thomas

Let's Review

Read: God's Food Resources

Fill in the blanks

1. God's _____ principle of _____ the _____ is not merely an expression of _____ or _____.

2. The obedience of this command brings the _____ _____ of the _____.

3. God's _____ are _____ to us and then we give them to the _____.

Read: Who May Want To Help

4. Who may want to help? (Give at least 5 answers)

   a. _____

   b. _____

   c. _____

   d. _____

   e. _____

5. Many _____ want to give, they just _____ know where the _____ is.

6. What is a joyful offering?

7. Gardens can be used to _____ any _____

8. Sometimes farmers or ranchers have surplus food. This may include:

   a. _____

   b. _____

c. _____

d. _____

e. _____

f. _____

9. This food may be _____-_____ _____, an over abundance of crops or after harvest _____.

10. It is important to _____ good _____ with those who give to you.

Read - Presenting The Need

Fill in the blanks

11. Visit local _____ and _____, and talk to the _____ or _____.

12. Assure them that absolutely nothing will be _____, _____, or _____.

13. It is important to :

a. _____

b. _____

c. _____

d. _____

14. It is important NOT to:

a. _____

b. _____

c. _____

d. _____

Read - The Daily Route

15. With God the _____ are _____.

16. In the morning before you start:

    a. _____

    b. _____

    c. _____

    d. _____

17. It is important not to be: _____

18. Be _____, make sure your appearance is _____ and
_____.

19. Always clean up _____ that you make and perhaps those that you
_____ _____.

20. Never allow _____, but keep _____ _____
about God's blessings.

Read – When the Truck Comes In

21. Show _____ sheet and all items received to those who have the
_____ over the _____ _____.

22. This will help them decide where to _____ the _____ and
where it will be _____.

23. A job _____ doing is worth _____ _____!

24. The entire _____ relies on _____ _____,
we must be _____.

25. _____ those who _____ food and those who _____
the food, to _____ damaged food items.

26. Use your _____ to life rather than your _____.

27. When the food is _____, pick up _____ food so
that _____ is wasted.

28. _____ all dairy products always stacking the _____ items
toward the front, ready for _____.

29. If you bring home _____, sort out the _____ ones.

30. Whatever _____ you make, _____ them up right away.

31. All food should be _____ so that the _____ items are in the _____ and easiest to reach.

32. Never _____ your job for _____ else to do.

Score exercises 1 – 32

Let's Review

DO NOT look back at the Training Pac while completing this Review.

True or False (Write "T" if the answer is True and "F" if the answer is False)

1. _____ God's kingdom principle of feeding the hungry is not merely an expression of sympathy or charity.

2. _____ The obedience of this command brings the acceptable year of the Lord.

3. _____ Many people want to give, they just know where the need is.

4. _____ Gardens can be used to supplement any food ministry.

5. _____ It is important to develop casual relationships with those who give to you.

6. _____ Visit long distance stores and businesses, and talk to the supervisors and employees.

7. _____ Assure them that everything will be sold, traded, or wasted.

8. _____ It is important to be friendly, smile.

9. _____ Go with an anointing knowing that God is sending you.

10. _____ With God the possibilities are limited.

11. _____ Do not beg or use slang language.

12. _____ Do not become discouraged when someone says no!

13. _____ Be casual, make sure you'll appear, wear T-shirt and jeans.

14. _____ Always clean up spills that you make but perhaps not the one that you didn't make.

15. _____ A job worth doing is worth doing right!

16. _____ You do not need to train those who give and receive food, just give the damaged items too.

17. _____ Use your back to lift rather than your knees.

18. _____ When stacking, the newest items should be towards the front and the oldest items towards the back.

19. _____ If you bring home eggs, sort out the cracked ones.

20. _____ Never leave your job for someone else to do.

Score exercises 1 – 20

Read – Handling and Storing Food

Fill in the blanks

1. When handling food it is important to:

   a. _____

   b. _____

   c. _____

2. Move the _____ potato chips to the _____, so they will be used first.

3. Mark _____ dates on all _____ items to _____

4. that they will be _____ out while they are still _____.

5. If _____ is _____ or _____ placed into _____ it will become _____.

6. The _____ of fruit should _____ _____ the _____ of the _____.

7. It is better to store _____ _____ with a _____ this will help when _____.

8. All food with _____ _____, such as onions and _____, should be kept _____.

9. _____ should be stored with stems _____.

10. If tomatoes are stored with stems up:

a. _____

b. _____

11. Stack boxes _____. Be _____ not to put them on

_____.

12. _____ each stack so that the _____ are less likely to

_____.

13. Canned items and dry goods should be stored:

a. _____

b. _____

c. _____

d. _____

Choose the best answer

14. Transfer dairy items, such as milk from a cracked or leaky jug into:

a. fresh, clean, sealable containers

b. another container.

15. Mark these containers with the correct date

a. so they are ready for distribution

b. so they are ready for storage

16. Gently handle each item that

a. will be given to those who have no need

b. will be given to those in need

Score exercises 1 – 16

Read – Personal Food Orders

Choose the best answer

17. Food areas should be kept

    a. by the dirt, germs and flies

    b. clean, away from dirt, germs and flies

18. Do not put food

    a. on the floor, sometimes observe health code laws.

    b. on the ground, always observe health code laws.

19. This food is a blessing

    a. from the Lord

    b. from rich people

20. When handling food it is important to

    a. rinse your hands with water

    b. wash your hands with soap and water

21. Fresh food is important

    a. in everyone's diet

    b. in no ones diet

22. Many dietary problems would be prevented

    a. or even cured by proper eating habits

    b. by eating unhealthy food

Fill in the blanks

23. How do we give God's food?

    a. _____

    b. _____

    c. _____

24. As you make the _____, ask the _____ to bless the
_____. Pray for the _____ who will receive the food.

25. This is not a _____, it is a _____ from the Lord!

26. Use the most _____ items first.

27. Give _____ meals from the _____ food groups.

28. Separate _____, food items that need to be

_____.

29. Place _____ in a box or bag with the _____ food items

at the bottom and the more _____ food items toward the top.

30. Arrange _____ order so that it has a _____ appearance.

31. If you are putting _____ _____ in the _____ of

someone's car, always _____ to make sure that there is nothing that might

_____ the food.

32. What can contaminate food?

a. _____

b. _____

c. _____

33. We are not _____ people to ourselves, we are drawing

_____ to _____ _____ by His

_____.

34. As you give _____ for God's _____ on the

_____.

35. God's _____ are an _____ of His love for _____.

Score exercises 16 – 35

Matching

1.___ When handling food it is important to	(a) exceed the height of the box
2.___ Move the older potato chips	(b) should be kept separate
3.___ If food is thrown or roughly placed into boxes it will become	(c) everyone's diet
4.___ The level of the fruit should not	(d) to the front
5.___ All food with strong odors like onions and broccoli	(e) the correct date
6.___ Stack boxes	(f) always wash your hands with soap and water
7.___ Mark the containers with	(g) neatly
8.___ Do not put food on the	(h) blessing from the Lord
9.___ Fresh food is important in	(i) ground
10.___ This is not a job, this is a	(j) damaged

Score exercises 1 – 10

Let's Review

Pre-Test

Choose the best answer

1. God's blessings are given to us and then we give them

    a. to others

    b. to ourselves

2. Many people want to give, they just don't know

    a. where the need is

    b. where to put the food

3. Gardens can be used to supplement

    a. any church organization

    b. any food ministry

4. It is important to develop good

    a. charisma to those who give to you

b. relationships to those who give to you

5. Assure them that nothing will be

    a. sold, traded or wasted

    b. bought, given or kept

6. It is important to be

    a. rude, unfriendly and frown.

    b. friendly, smile. The joy of the lord is with you.

7. With God the possibilities are

    a. limited

    b. unlimited

8. It is important not to be

    a. late for any appointment

    b. to take your time getting there

9. Never allow complaining, but keep

    a. good attitudes about God's blessings

    b. good attention to God's blessings.

10. A job worth doing is

    a. worth doing wrong

    b. worth doing right

True or False (Write "T" if the answer is True and "F" if the answer is False)

11. _____ Always leave your job for someone else to do.

12. _____ When handling food it is alright to keep touching your hair and mouth.

13. _____ Move the older potatoes to the front so they will be used first.

14. _____ If food is thrown or roughly placed into boxes it will become damaged.

15. _____ The level of fruit of fruit should not exceed the height of the box.

16. _____ All food with strong odors like onions and broccoli can be kept together with other foods.

17. _____ Put the boxes on of each other without interlocking them.

18. _____ Transfer dairy items from a cracked or leaky jug into any container.

19. _____ When handling food it is important to wash your hands with soap and water.

20. _____ This is not a job, it is a blessing from the Lord.

Matching

21. ___ Food can be contaminated by	(b) pray for God's blessing on the food
22. ___ Arrange the food order so that it has a	(a) Jesus Christ
23. ___ We are not drawing people to ourselves we are drawing people to	(c) pleasant appearance
24. ___ As you give	(d) expression of His love for man
25. ___ God's gift are an	(e) bleach, fuel, detergent

Score exercises 1 – 25

Let's Review Key

God's Food Resources

1. kingdom… feeding…hungry…sympathy… charity

2. acceptable… year…Lord

3. blessings… given…people

Who May Want To Help

4.

    a. your church

    b. family and friends

    c. farmers and ranchers

    d. owners and managers

    e. grocery stores or cold storage facilities or rice, flour, grain companies or large corporations or small businesses

5. people… don't…need

6. a day set aside by your church to collect food that will be given to those in need.

7. Supplement…food…ministry

8.

    a. livestock

    b. poultry

    c. eggs

    d. grains

    e. dairy

    f. produce

9. non-marketable…culls…gleaning

10. develop…relationships

Presenting The Need

11. stores…businesses…owners…managers

12. sold…traded…wasted

13.

   a. Carry a letterhead showing that you are helping those who have a real need.

   b. Be friendly, smile. The joy of the Lord is with you.

   c. When possible, take someone with you. Let your words uplift and encourage one another. Pray together or guidance and pray for the people you will come in contact with that day.

   d. Go with an anointing, knowing that God is sending you.

14.

   a. give a careless presentation

   b. wear untidy, torn or dirty clothes

   c. do not beg or use slang language

   d. do not become discouraged when someone says no!

The Daily Route

15. possibilities… unlimited

16.

   a. gather boxes you will need for the day

   b. check the oil in your vehicle

   c. check the gas

   d. write down the mileage

17. late for any appointment

18. courteous…clean…neat

19. spills…didn't…make

20. complaining…good…attitudes

When the Truck Comes In

21. inventory…responsibility…food… ministry

22. store… food…distributed

23. worth… doing…right

24. provision… God's…blessings…grateful

25. Train…give…receive…damaged

26. knees… back

27. re-boxed… remaining…nothing

28. Separate…oldest…distribution

29. eggs…cracked

30. spills…clean

31. organized…oldest…front

32. leave…someone

Let's Review

1. T

2. T

3. F

4. T

5. F

6. F

7. F

8. T

9. T

10. F

11. T

12. T

13. F

14. F

15. T

16. F

17. F

18. F

19. T

20. T

Handling and Storing Food

1.

    a. to keep your hands away from your hair and mouth

    b. use disposable gloves when necessary

    c. always wash your hands with soap and water before handling food

2. older…front…first

3. code…incoming…ensure…given…fresh

4. food…thrown…roughly…boxes…damaged

5. level…not…exceed…height…box

6. fragile…food…lid…stacking

7. strong…odors…broccoli…separate

8. Tomatoes…stems

9.

    a. moisture collects at the stem and causes mold to grow

    b. the stems will puncture the tomatoes stored on top of them

10. neatly…careful…walkways

11. Interlock…boxes…full

12.

    a. in a cool dry place

    b. off the ground

    c. away from mice

    d. and in air tight containers

13. a

14. a

15. b

Personal Food Orders

16. b

17. b

18. a

19. b

20. a

21. a

22.

    a. How much food do we have?

    b. How many people will we feed?

    c. Was this food gathered for this group or are other groups coming at a later time?

    d. How long will this food last?

23. order…Lord…people

24. job…blessing

25. perishable

26. balanced…four

27. produce…refrigerated

28. produce…sturdier…fragile

29. food…pleasant

30. food…order…trunk…check…contaminate

31.

    a. stored fuel

    b. detergent

    c. bleach

32. drawing...people...Jesus...Christ...love

33. pray...blessings...food

34. gifts...expression...man

1. f

2. d

3. j

4. a

5. b

6. g

7. e

8. i

9. c

10. h

Pre-Test

1. a

2. a

3. b

4. b

5. a

6. b

7. b

8. a

9. a

10. b

11. F

12. F

13. T

14. T

15. T

16. F

17. F

18. F

19. T

20. T

21. e

22. c

23. a

24. b

25. d

Final Test

True or False (Write "T" if the statement is True and "F" if the statement is False) 1.

1. _____ Transfer dairy items from a cracked or leaky jug into any container.

2. _____ The level of fruit should not exceed the height of the box.

3. _____ Move the older potatoes to the front so they will be used first.

4. _____ This is not a job, it is a blessing from the Lord.

5. _____ When handling food it is alright to keep touching your hair and mouth.

6. _____ If food is thrown or roughly placed into boxes it will become damaged.

7. _____ When handling food it is important to wash your hands with soap and water.

8. _____ Put the boxes on top of each other without interlocking them.

9. _____ Always leave your job for someone else to do.

10. _____ All food with strong odors like onions and broccoli can be kept together with other foods.

Choose the best answer

11. God's blessings are given to us and then we give them

   a. to others

   b. to the trash

12. It is important to develop good

   a. hidden agenda to those who give to you

   a. relationships to those who give to you

13. It is important to be

   a. rude, unfriendly and frown. You just had a bad day.

b. friendly, smile. The joy of the lord is with you.

14. Never allow complaining, but keep

   a. good attitudes about God's blessings

   b. focusing on the bad things in life.

15. With God the possibilities are

   a. limited

   b. unlimited

16. Assure them that nothing will be

   a. sold, traded or wasted

   b. given to those in need

17. A job worth doing is

   a. worth doing wrong

   b. worth doing right

18. Many people want to give, they just don't know

   a. where the need is

   b. want to be lazy

19. Arrange the food order so that it has

   a. a pleasant appearance

   b. an alphabetical arrangement

20. Gardens can be used to supplement

   a. any church organization and food ministry

   b. those who won't eat vegetables

Flinal Test Key

True or False

1. F

2. T

3. T

4. T

5. F

6. T

7. T

8. F

9. F

10. F

Choose the best answer

11. a

12. b

13. b

14. a

15. b

16. a

17. b

18. a

19. a

20. b

CHAPTER 2

Close Together Planting

Raised-bed gardening is a form of gardening in which the soil is formed in three-to-four-foot-wide (1.0–1.2 m) beds. The soil is raised above the surrounding soil and is sometimes enclosed by a frame generally made of wood, rock, or concrete blocks, and may be enriched with compost.

The vegetable plants are spaced much closer together than in conventional row gardening. The spacing is such that when the vegetables are fully grown, their leaves just barely touch each other, creating a microclimate in which weed growth is suppressed and moisture is conserved. The close plant spacing and the use of compost generally result in higher yields with raised beds in comparison to

Plant Growth

Flowers make more plants by forming seeds. Fruit are natural seed holders.

Leaves make food. Sunlight works in the leaves to take raw materials from soils and air

Leaves "breathe".They draw in carbon dioxide and let out oxygen which we breathe) and water vapor.

Stems are plumbing

Nutrients and water are taken from the soil through the roots and are drawn up the stems to the leaves. Then,food from the leaves continues through the plant

Main Roots anchor and hold the plant upright against wind and weather.

Tiny Root Tips (many times, so small you can't see them) draw water and nutrients through fine tips only.

Soil needs air and water, and needs to be "loose" and easy to dig and plant in.

New Word:

Screen - Wire mesh fastened to a frame placed at an angle - (see picture).

To sift soil, throw sand, stones, grass against mesh. San d will go through, stones

and grass will remain behind. When you get ready to plant a garden, you must first decide where to plant. You need to think of the following to make sure your garden can grow well.

1. **Water**: You will want to be able to water your garden. Plant your garden somewhere close to a well or stream or some other source of water. Vegetables need a lot of water. A good way to decide where your garden should be and how large is to let your water supply provide the size.

2. **Sun:** Most vegetables grow best in direct sunlight. Pick the sunniest spot available because your plants have to have a minimum of 6 hours of a day of sunshine.

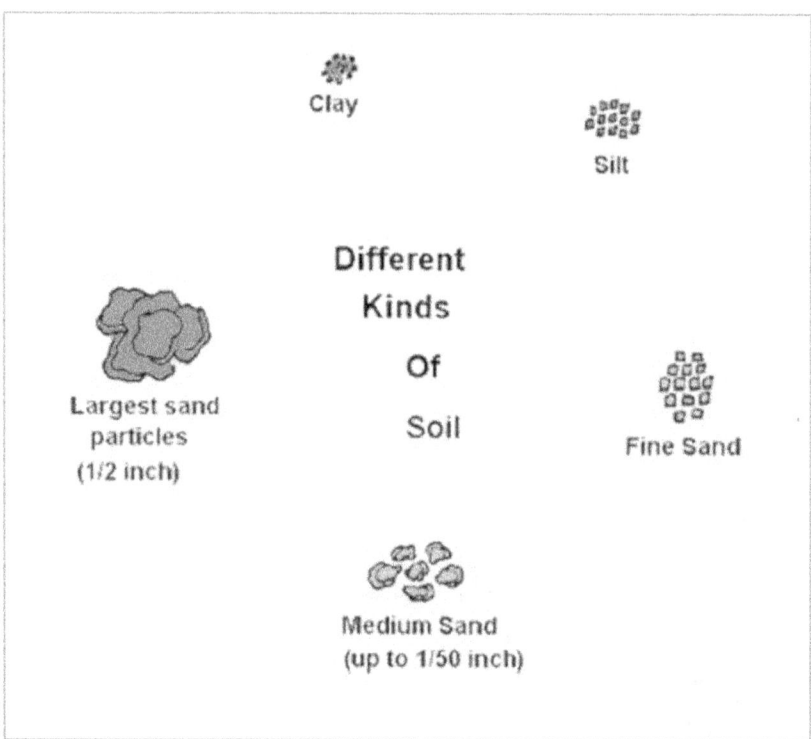

3. **Soil**: You must look at the soil where you want to put a garden. Sometimes soil is very hard or mostly rocky. If the soil is too bad, you may not be able to plant in it. Some very hard soil or rocky soil can be chopped, dug and sifted to make it soft enough for plants. To sift it, put it through a screen or very small netting to take out the rocks and big hard clods. If the soil is not right, it cannot grow plants.

Good soil is land used for planting. Good soil is a mixture of clay, sand and rotted plants. It also has small living animals which are so tiny you cannot see them with your eyes.

Soil is different in different parts of the world.

One of the most important things about gardens is to make the soil just right for your plants to grow. It is important to know what kind of soil you have.

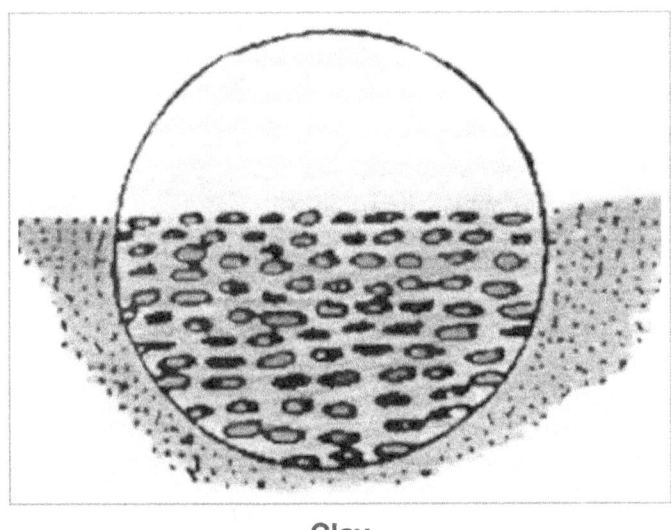

Clay

Clay is made up of very tiny pieces of soil. (It can be used to make pottery and bricks when it is dried.) These tiny pieces are flattened and fit closely together so tightly that it is hard for water and air to get inside.

When soil has too much clay in it, the plant roots cannot grow deep into the soil because it is too hard. Air does not get to the roots and water runs off the top; when clay gets wet it dries out slowly because water goes down slowly. But clay has a lot of very important nutrients in it, so it is good to have some in your soil. Too much clay is not good because it is too hard.

Sandy soil has larger particles than clay, and they are more rounded, rather than flat. This shape allows more room between each particle so air and water can flow through it more easily.

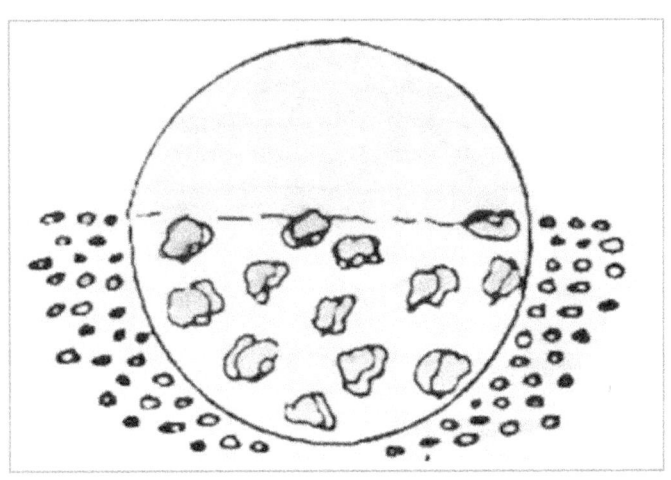

Sandy Soil

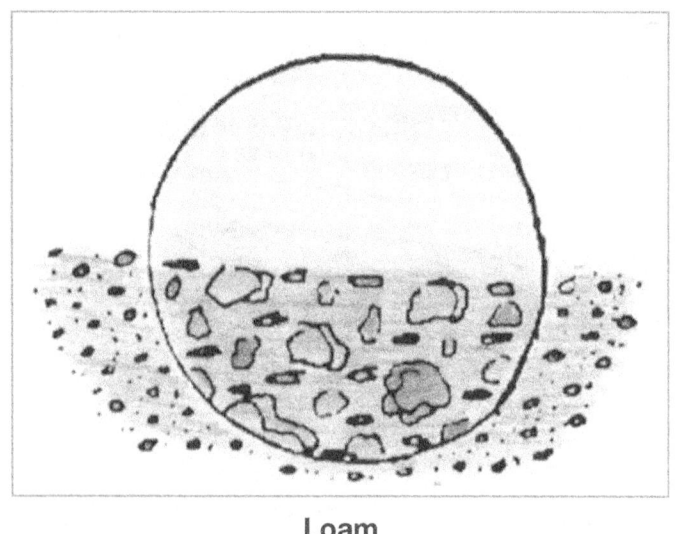
Loam

It also dries out more quickly so it must be watered a lot and too much watering can wash away valuable nutrients.

Loam is best for growing a garden, it is a mixture of clay, sand and compost. Compost is soft, crumbly, brownish or blackish and is made by allowing dead leaves, plants and waste to rot. It is made by man to help in the garden. Loam is ideal for planting, because it has clay, with its wet, healthy nutrients and sand, to drain well and let water and air go down to the roots of the plant and compost to help feed the plants.

Test your soil:

Make a ball with a handful of soil. Throw the ball up into the air. If the ball sticks together before it lands, there is too much clay in the soil. You will have to add some sand.

Another way to test the soil is to water it. If the water does not sink into the soil, runs off or form puddles on the surface, there is probably too much clay in the soil.

Before you start your garden, remove all large rocks from where your garden bed will be; remove all grass; remove all weeds. If the weeds were very thick and were growing everywhere, plant legumes the first year. Legumes are beans or peanuts. These will help make your soil better the next year.

Close Together Planting

Close together planting is a way of gardening first used by the French farmers centuries ago. It is good especially if you have only a very little space to plant in.

The farmers who started this kind of gardening had lots of manure. If you have lots of manure cover your bed-to-be with six inches of good rotted compost or manure before you ever start your deep digging. Then mix in more compost or manure as you go. (Remember, you can only do this if you have a lot of manure).

The way you plant your garden depends upon the kind of weather you have in your area. If you live in a dry place where it is very hard to get water, you would want to plant your garden so that you catch every drop of water you can on your plants. If you live in a place where there are monsoons and heavy rains you would want to plant your garden so that the plants are raised higher and will not be washed away.

If your climate is dry, you will want to make rows. You will want to grow your vegetables down inside the trenches. This way you catch every drop of water on your plants. You can also make your garden lower than the ground around it. Make a high edge all the way around. Then, flood the lowered bed with water. This will keep all the water on your plants.

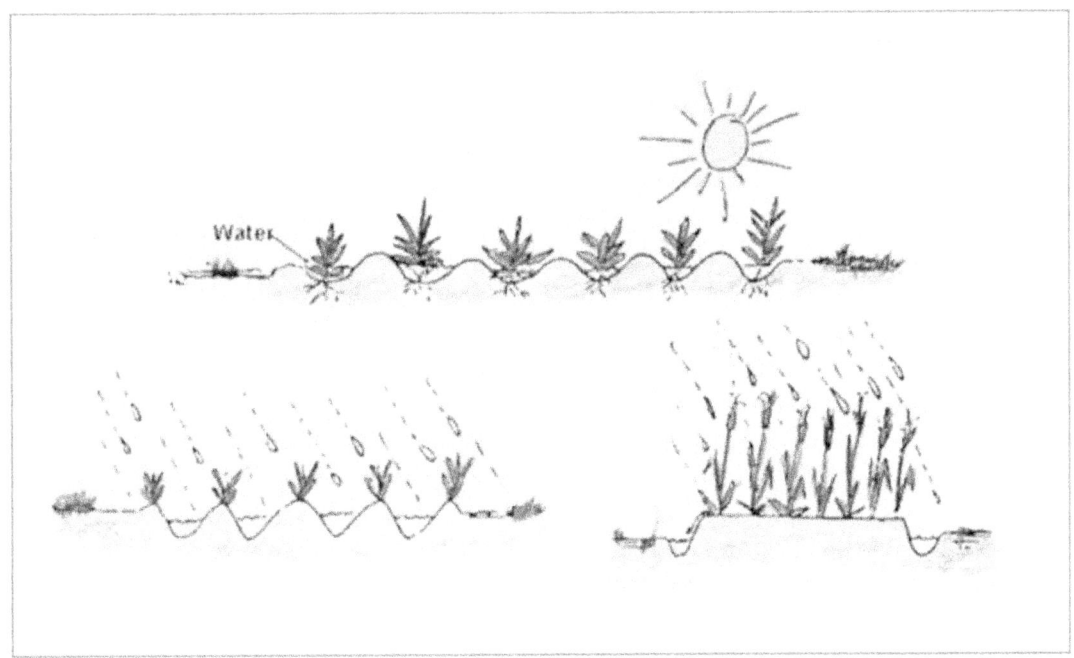

If your garden is in a wet area, you will want to plant you garden on raised beds. This way the plants are above the water and will not be washed away.

Step One: Make a Garden Bed by Deep (double) Digging

1. Dig a trench 8 foot long and a foot wide by removing the top nine inches of soil alongside

2. Loosen the soil in the trench another nine inches. The trench should then be 18 inches deep, half filled with loose soil. REMEMBER TO ADD SAND IF THERE IS TOO MUCH CLAY IN YOUR GARDEN. Twelve inches is deep enough.

3. Cover the bottom nine inches of loose soil in the first trench with nine inches of top soil from a second trench dug alongside the first.

4. Loosen the bottom nine inches of the second trench.

5. Continue these first four steps until you have dug three or four trenches right alongside each other.

6. Cover the bottom nine inches of soil in the last trench with the nine inches of top soil that you had removed from the first trench.

7. Try to get the soil broken down to a fine texture.

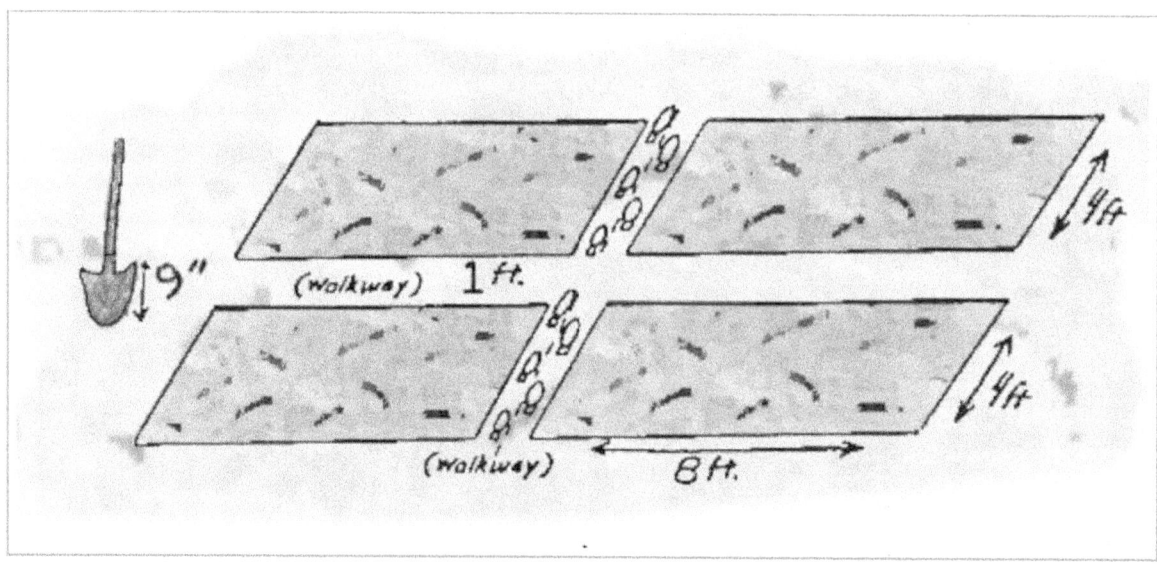

Level the garden bed so it is completely flat and level. Try to break down the soil to a fine texture.Then raise the garden bed four to six inches above the ground around it. This will keep it from being flooded by heavy rains.

The end result, is a raised garden bed 8 foot long and 4 foot wide. Four feet is a good width for a garden bed because you can easily reach into the middle of it to plant, weed or harvest without stepping into it.

You can make as many of these beds as you would like. Leave one foot walkway between each bed. If you do not have enough space, you can make the bed smaller.

Benefits of a Raised Garden Bed

1. The roots of the plants will grow deep into the soil. Deeply rooted plants have healthier roots and stand stronger.

2. Water sinks down deep into the soil and stays there.

3. Air can get to the roots and aid the growth process.

4. Compost is mixed into the soil helping the plants get minerals they need and providing nutrients at all levels.

Step Two: Your Fertilizing Process

Sprinkle a dusting bone meal, some wood ashes an inch or more of well rotted manure (if you did not use manure when you made your garden bed.) over the top. Rake or cultivate this fertilizer into the top 3 to 6 inches.

Step Three: After Your Deep Digging is Finished

1. Soak the soil with gentle spray of water.

2. Your garden bed should be smooth and flat, but soft.

3. Leave this finished bed for two days.

Note: If there is a shortage of water, several thickness of newspaper soaked good and laid around the base of plants will keep them moist for several days.

1. PREPARE THE SOIL

2. ADD COMPOST OR MANURE

3. SPADE IT IN

4. ADD BONE MEAL, WOOD ASH AND OTHER ADDITIVES

5. RAKE SMOOTH.

The Planting Process

1. Take an inch thick stick and draw rows lengthwise in the prepared garden bed.

2. You will then have hollows and ridges.

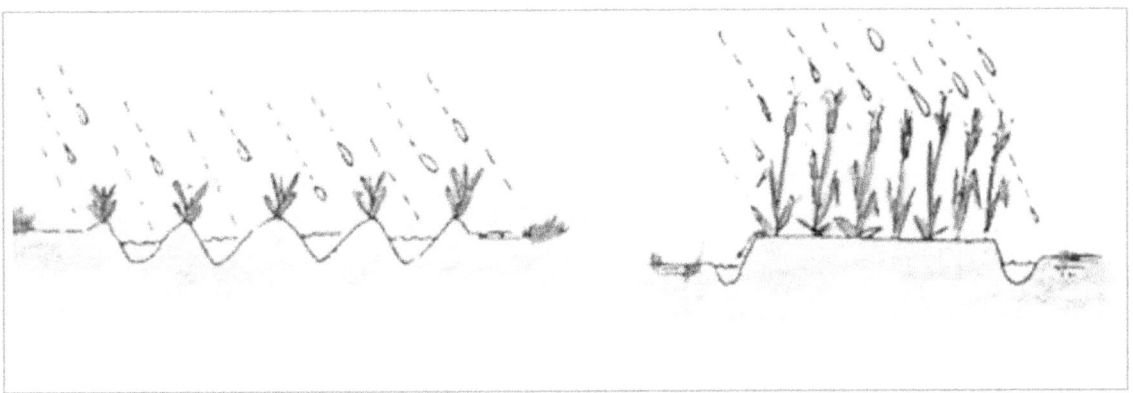

   a. If you have a lot of rainfall, plant seeds in rows along the ridges.

   b. If you have little rainfall, plant seeds in hollows.

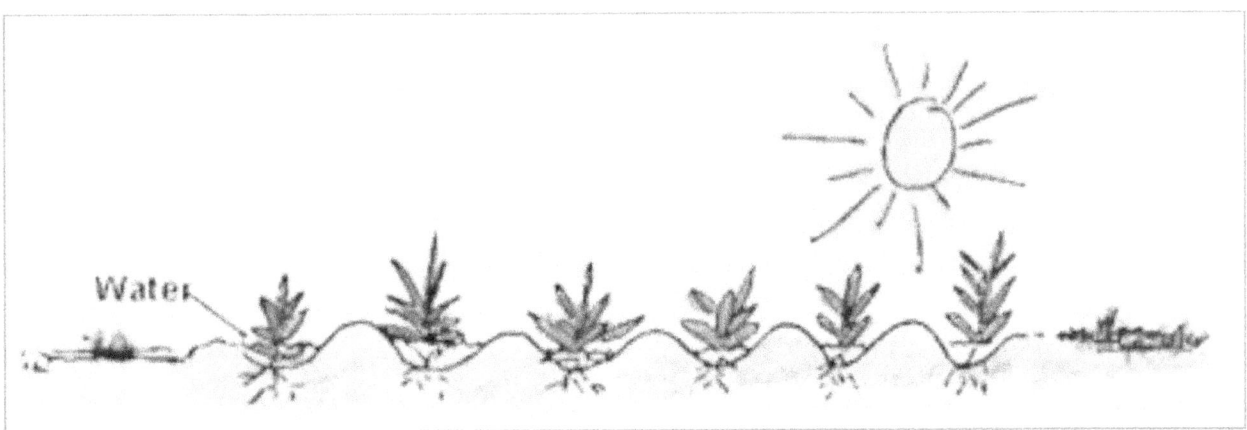

It is possible to plant the seeds close together because the garden bed has been prepared by deep digging. This allows the roots of the plants to grow deep. Since the roots grow deep they get plenty of moisture and nutrients from the soil and the plants can grow close together. This will help your plants to produce many large and tasty fruits, vegetables and melons.

In close together planting, the plants are planted so close that the outer leaves of the plants will touch as they get bigger. Whether you are setting out transplants or planting from seed, you are going to put the plants closer than you ever have before. When the plants are big, you will have a solid garden of plants in your little space. If your soil is deep enough; is fertilized well; is watered well and you keep it weeded carefully, that little piece of earth will be able to support all those plants.

When the plants are fully grown, you should be able to see the ground beneath. This kind of shading helps the soil stay moist and protected from too much wind and rain. The plants will also protect each other from too much wind and rain. Weeds do not grow well because they are shaded too much by the good plants.

Plant	Distance Apart (in inches)	Depth in Soil (in inches)
Corn	8"	2"
Okra	8"	1"
Tomatoes	24"	0.5"
Bush Beans	4"	1"
Pole Beans	8"	1"
Peppers	18"	1"
Jicama	4"	1"
Squash	18"	1"
Pumpkins	18"	1"
Melons	18"	1"
Cucumbers	4"	1"
Onions	1"	1"
Carrots	1"	1"
Radishes	0.5"	0.25"

Close together planting like this works better with leaf and root crops like lettuce, spinach, cabbage, beets, carrots and turnips.

Big vegetables like corn and squash are going to take more room no matter how hard you try.

Look at the chart to see how far apart to plant your seeds. The plants are listed from shortest to tallest (unless they are put in a trellis).

Carefully follow planting instructions on seed packet for each type of seed.

Compost, Fertilizer, and Soil Additions

III. COMPOST, FERTILIZER AND SOIL ADDITIONS

New Words:

Waste - paper, newspaper, cardboard (not tins or plastic).

Organic Material - vegetable leaves, and foodstuff which cannot be eaten. Also tea and coffee grounds.

Compost is rotted leaves, plants and dead insects. It is important because it helps the soil grow plants better. You should never use Citrus fruit or the peelings from them, nor cabbage leaves in compost. Remember that the best soil is:

1/3 clay

1/3 sand

1/3 compost

There are Three Kinds of Compost

1. Manure

2. Rotted plant and animal life. Eggshells are excellent for compost.

3. Earthworm castings and rotted root hairs (very small roots.)

Manure is very good compost. It has already been changed by being chewed into pieces by some animal. Dry manure is compost. Fresh manure is just manure. It contains worms and diseases. Do not use manure that has not been dried. Do not use human waste -- Human waste carries many diseases.

The best manure for your garden is horse manure. Remember not to use any fresh manure. Get some manure that has been completely dried. It should be at least two weeks old.

Cow manure is good also. You can use a lot of it in your garden. Rotted cow manure ruins beans. Beans do not need as much fertilizer.

Sheep or goat manure is not so good for your garden unless you use only a little. Sheep and goat manure is very strong. If you use too much of it, it will be too strong for the plants and will kill them.

Chicken manure is also strong but you can use it if you mix it with sawdust and then wait for a few weeks before you use it.

Humus in the soil helps beat down the aphid population and also helps so much when there is a drought.

Garden residues for compost - Tomato and squash vines, flower stems and pepper plants. Grass cuttings are excellent also.

Compost Does A Lot for Your Garden

New Word:

Recycle - to treat something or allow it to rot in order to use it again.

When you add compost to the soil, you are improving the soil, making it easier to work with. Water and air penetrate better. Also, compost helps recycle waste, leaves, grass and other organic material. Compost allows the plants to choose the nutrients they want.

Compost also makes the fruits and vegetables taste better.

Compost is made from plant materials, (such as leaves, twigs, and grass), garbage and manure. There are several ways of making compost.

1. Use a barrel. (This method takes about two weeks.)

2. Make a pile. (This method takes about three months.)

How to Use a Barrel:

1. Cut out the top and bottom of a rain barrel or an oil drum.

2. Fill the drum with 1/3 plant material, 1/3 waste and 1/3 manure. Chopped into very small pieces and mixed well.

3. Cover everything in the drum with one inch of dirt to prevent odors and flies. This is very important.

4. Help the compost rot by making it moist.

You will have to turn the contents of the barrel once every three days. Do this by lifting the barrel off the material, mix the material and put it back in the barrel.

After two weeks you will have a barrel full of compost.

How to Make a Pile:

Find a place where water drains off quickly and make a compost pile there.

1. Begin by loosening the soil at the place where you are going to make your compost pile.

2. Get some twigs and lay them across each other until you have a pile about three inches high. This is to allow air to circulate = ventilation.

3. Put six inches of plant material, (grass, leaves, and small twigs), on the pile of large twigs.

4. Add four inches of waste.

5. Cover the whole pile with four inches of fresh manure. (Manure is the most important ingredient of a compost pile. The manure will rot the rest of the material in the compost pile.)

6. Lastly, cover everything with one inch of dirt to prevent odors and flies. (This is very important because flies carry many diseases.)

7. You can add layers of plant material, waste and manure (in that order) until the pile is five feet high.

8. When the pile is five feet high you should not add anymore plant material or waste to it. You can start a new compost pile.

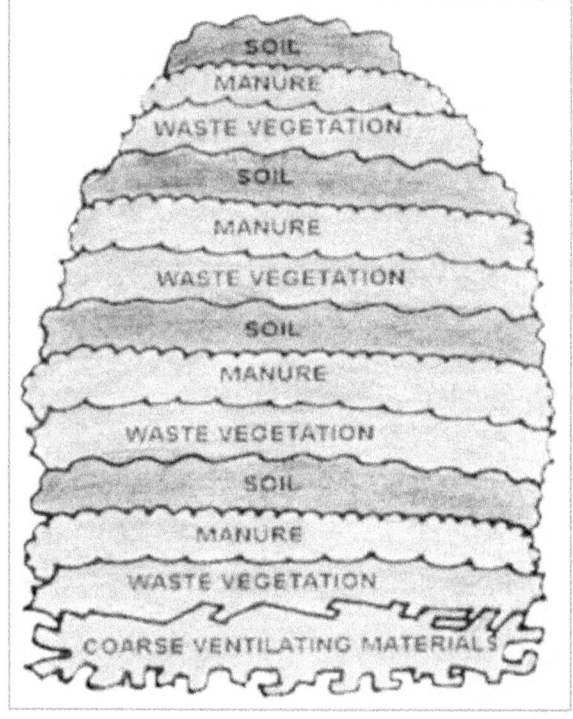

9. As it rots, the pile will shrink down to about two feet.

10. It is helpful to keep the whole pile moist while it is rotting.

After you have made this pile you still have not made compost. The pile will not become compost until it "rots".

You will have to completely "turn" or "mix" this pile about once a week with a shovel, pitch fork or rake. Take some of the material from the bottom of the pile and place it on the top of the pile. Keep doing this until the entire pile has been turned. At this time you may add more manure to the pile with dirt again, and wet it down again.

After about three months, (if weather is warm), all of the material will be rotted and you will have compost.

Since a pile of compost material takes about three months to rot you might want to start a new compost pile every month. That way you will always have some compost to put in your garden.

Soil Additions

Besides compost, there are some other things you can put in your soil to improve it.

Add limestone if the soil is too acid. (Soils in areas of heavy rainfall tend to be acid).

Add crushed rock or eggshells if the soil is too alkaline. (Areas of light rainfall tend to be alkaline).

Phosphorous, potassium, and nitrogen are important nutrients in the soil.

Grind up some bones for phosphorous.

Add wood ash for potassium.

Add fish, blood, or ground up horn for nitrogen.

Manure Tea for Feeding Plants

Two shovels of manure (cow or horse) in a 5 gallon bucket of water. Mix well, let it sit until the sediment settles to the bottom of the bucket, then dip out liquid and pour around the base of the plants.

Plan Your Garden

Planning Your Garden

Planning your garden has a lot to do with where you live. It has to do with how hot the sun is in your area. When you plant your garden make sure your plants will get enough sun but not too much sun. Too much sun will burn your plants.

If you live where the sun is very hot, you need to plant the taller plants toward where the sun rises, then the tall plants will help shade the shorter plants.

If you live where the sun is not very hot, you need to plants the shorter plants toward where the sun rises, then the shorter plants will get enough sun, and will not be shaded by the taller plants. The chart from Section 2 lists the plants from tallest to shortest.

Some plants like corn need the hot sun.

Other plants like lettuce need to be partially shaded.

Tomatoes, cucumbers, pole beans, and even pumpkins and squash can all be grown on trellises to save room in the garden.

Here is a picture of three kinds of trellises:

Corn, okra, pumpkins, and squash take up a lot of room. They must be planted together in a separate field.

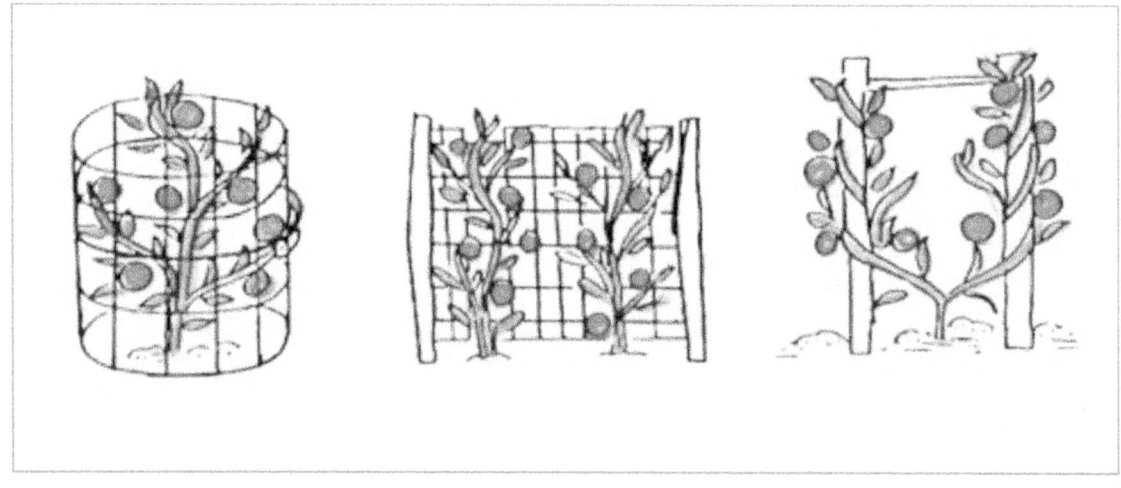

Peppers, tomatoes, herbs, and even carrots can be used as ornamental plants. They can be used as borders or hedges around your house or compound.

Remember to keep the garden bed moist at all times while the seeds are beginning to grow. It is best to water the garden lightly just after sundown or just before sunrise. Water once a day until the plants are three weeks old. When the plants are three weeks old, you will only need to water them every three days, but you will not water them lightly -- instead you will water them until the garden bed is thoroughly soaked. This heavy watering will allow the water to go down deep into the soil. The plant roots will be encouraged to grow deep to get the water. The plants will become stronger.

Rain is God's way of watering a garden. Rain falls as little droplets which do not disturb or wash away the soil. Rain water is good water which is free of things that can hurt your plants. This makes rainwater better for watering a garden than river water. Water is not the only thing that falls to the ground during a rainstorm. Large amounts of nitrogen also fall to the ground with the rain water. Nitrogen is important for plant growth.

If your garden is in a wet area, you will want to plant your garden on "raised beds". This way, the plants are above the water and will not be washed away.

Plant by Phase of the Moon

The pull of the moon's gravity will help pull the seeds out of their shells. Plant leafy crops in the light of the moon. Moonlight helps the leaves grow. Planting root crops in the dark of the moon will help root growth. This does not mean to plant your crops at night it means to plant during the days of the moon phase.

Starting Seeds in Flats

New Words:

Flat - A flat is a box with very low sides to grow you seedlings in until they are just a few inches tall.

To Thin - carefully pull out the extra weaker plants to allow the stronger plants to grow evenly.

Sometimes it is better to plant certain seeds in flats and transplant them to the garden after a few weeks. This is especially important in cold areas. You can start your planting while there is still frost and cold, and transplant into the garden after the ground is warm enough. This way, you can start your garden early.

1. Seeds started in flats can be completely protected from the weather, insects and weeds until they are strong enough to be transplanted into the garden.

2. Because they can be completely protected, seeds grown in flats need less care and watering than seeds planted in the garden.

How to Start Seeds in a Flat

Remember be best soil is 1/3 clay, 1/3 sand and 1/3 compost.

1. Before mixing for a flat, mix the clay and the sand and bake in the oven at a medium temperature (300°F). This will kill all weeds and their seeds.

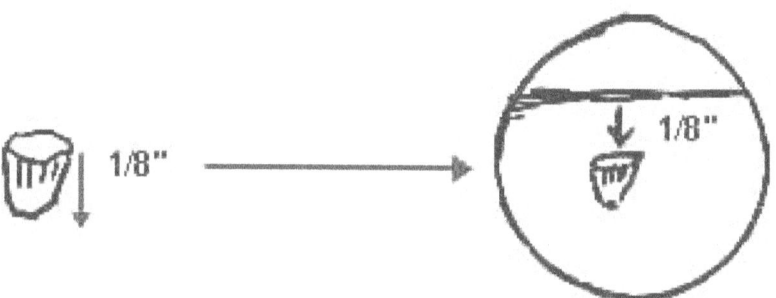

2. Add compost or dry manure to the clay/sand mixture.

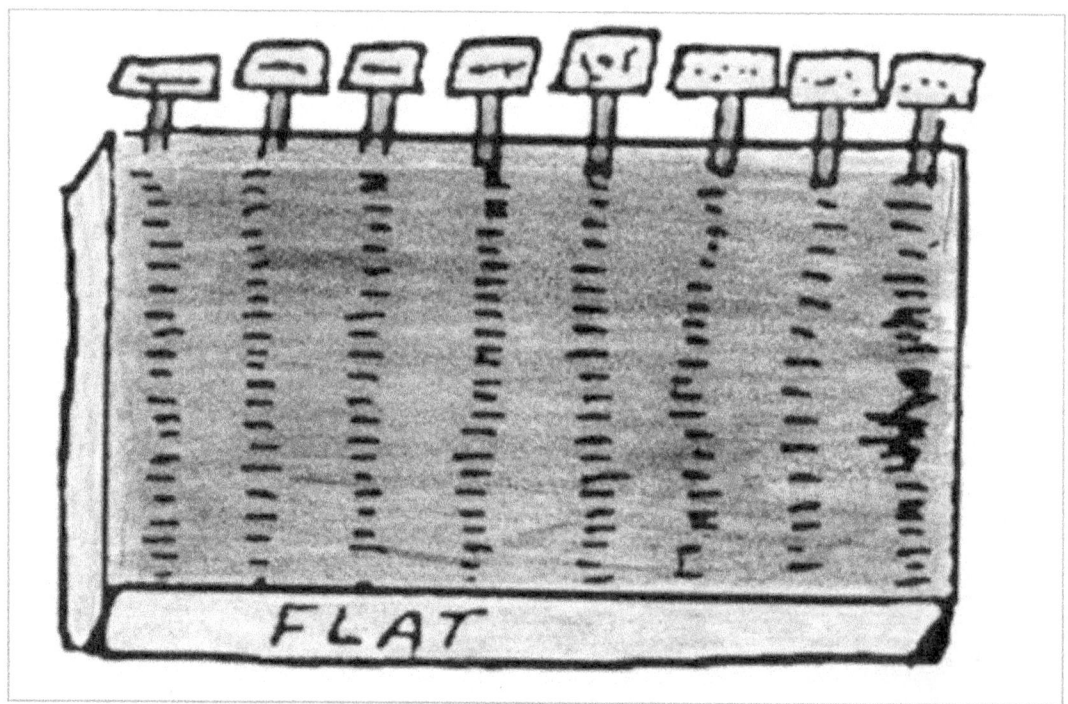

FLAT

3. Plant the seeds the same depth as the length of the seed. This is a good rule for all seeds if you do not know how deep to plant a seed.

4. Plant the seeds in rows.

5. Mark the rows so that you will know what is planted in them.

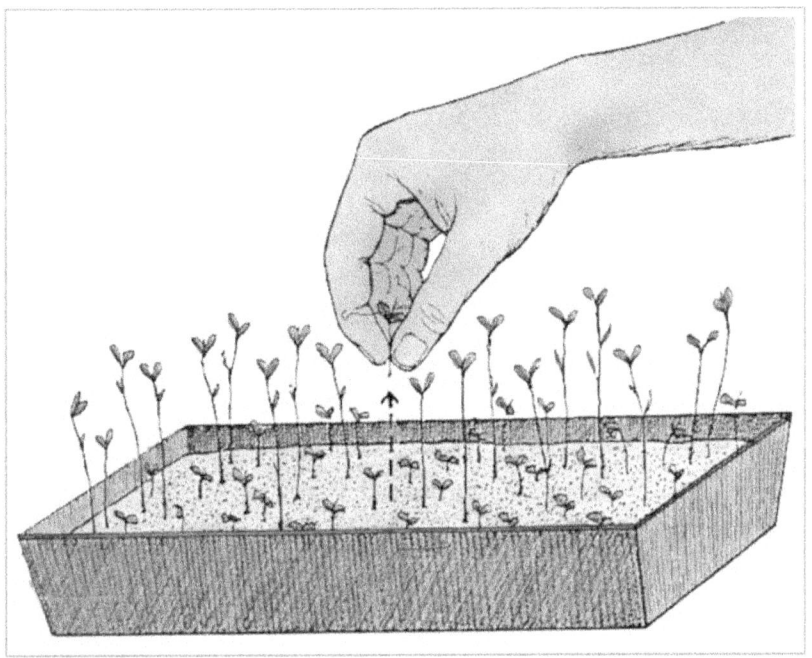

6. Keep the soil moist.

7. After the seeds have begun to grow, thin the plants so that they are not crowded.

8. When the second set of leaves begin to appear, the plants can be transplanted. (Or you can wait until the plants are three inches high.)

Transplanting

1. Prepare rows in the same way as you would for planting seeds (see The Planting Process).

2. Transplant late in the afternoon so that the plants will get some sun but not too much.

3.The soil in the flat must be moist. Gently lift each plant out of the flat keeping some soil around the roots. Try to keep the roots covered

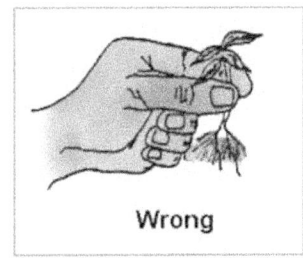

Wrong

4. Hold the plant very gently in your hand. Hold it by cupping your hand around the soil, never by the stems. The stems are very fragile and break easily.

5. In the row, make a hole deep enough and wide enough for your plant, using stick or a small trowel.

6. Gently place the plant in the hole, so that the first set of leaves is just above the

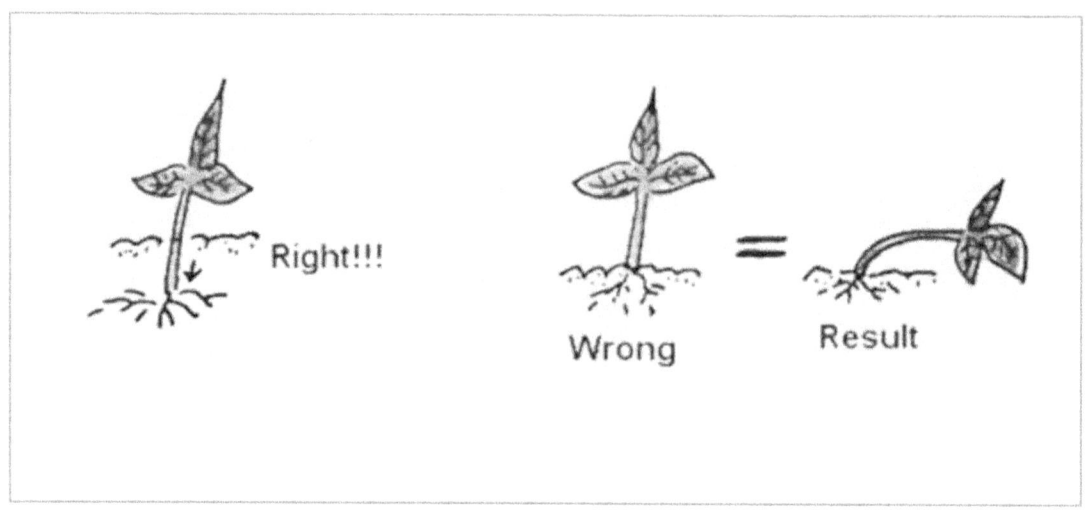

ground.

7. Press the soil down firmly but not tightly. (Lightly pat down the soil) After planting, firm the soil slightly with your hands to remove any air pockets. Don't wait for the newly transplants to show signs of wilt before watering again. A

little extra attention in the first few days after planting will ensure healthy plants.

8. When you have transplanted all the plants, water the bed gently but thoroughly.

    a. When many plants are grown together in a single flat, their roots

intermingle. The individual plants will be damaged less if you pull them apart with your hands rather than using a knife or other sharp instrument.

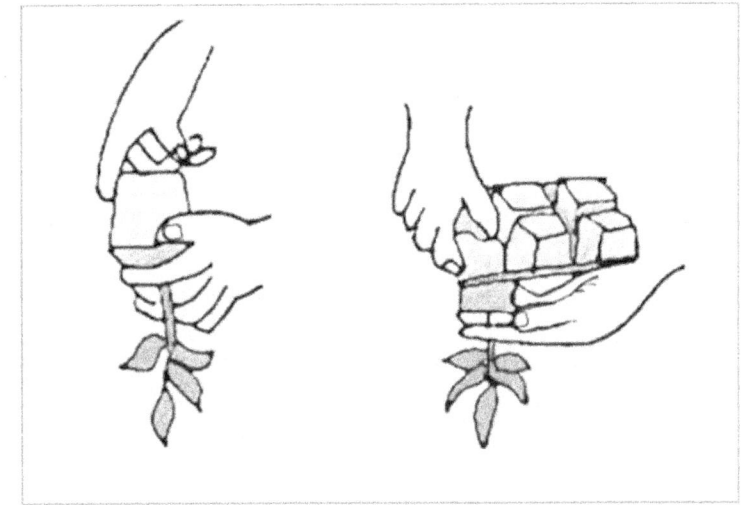

For transplants grown in individual plastic pots, tip the pot and tap the plant into your hand -- don't pull it out. Plants in six packs should be turned over and pushed out from the bottom of your thumb. Hold the soil in place with your other hand.

Watering your garden.

1. Here are two ways that you can catch rain water and use it on your garden.

   a) A Tank

   b) A Rain Barrel

2. Keep the garden moist at all times when the plants are first beginning to grow.

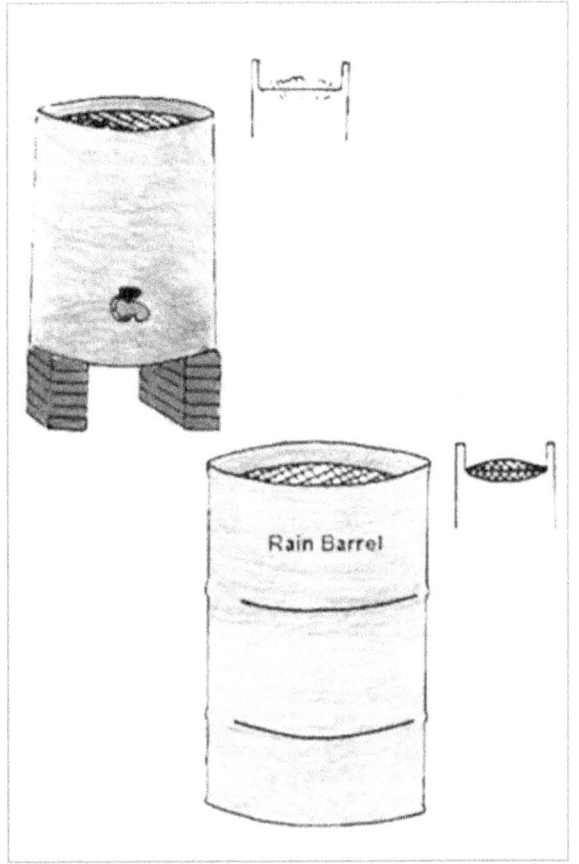

3. Use a watering can or spray nozzle on a hose to water like soft raindrops.

4. When the plants are three weeks old you can water more heavily but less often. Every 3 to 4 days, either from rain or by watering.

5. Cabbage, cauliflower, and broccoli like water on their leaves, and are lovers of cool weather. Radishes also do well in early spring. Most other plants do not like water on their leaves.

New Word:

Mulching - covering soil around the plants in order to keep moisture in and prevent weed growth.

6.   Mulching keeps your plants moist and helps keep weeds away. Use dried grass, leaves, manure, compost or sawdust as a mulch. Spread an inch or so of the mulch on the garden bed. Black plastic, brown paper, aluminum coated plastic and foil can also be used instead of mulch. You can determine the way you will mulch by your climate - for example - if you live in a very hot climate black plastic will burn your plants.

Vegetables need a lot of water. Flowers can survive longer without water than vegetables, so can trees and bushes. Roses and Chrysanthemums love a lot of water. Water often. Some vegetables never recover from a drought and almost all of them will

produce a lot more with abundant water. A good way to decide where your garden should be and how large it should be is to let your water supply decide the size. Unless you live in an area with dependable spring to fall, long abundant rains, don't cultivate and plan to garden in places that you can't get water to. If you are working with a limited water supply, figure that your garden needs at least an inch of water a week, either from the sky or from your water system.

Plant Mainenance

Irrigation and Water Systems

Soaker Hose

1. Take a hose and block it up at one end.

2. Make tiny holes evenly spaced along the length of the hose.

3. Attach the open end to your water source.

The water will spray up to 3 to 5 feet through the holes, depending on the water pressure. It has a gentle soaking action that is good for your tender plants and older ones too. If you have a big garden you can hook up to 2 or more lengths of hose.

Pictured here is an example of a soaker hose clean bucket or barrel in a high place about 10' high in or near a garden. The hose is attached to the bucket or barrel. Then, with a nail or a pin, a hole is made where each plant is located.

Bucket - Another way to water is to carry the water in buckets and pour it beside the plants. Some farmers have made very good gardens watering this way. Remember to be very gentle on young tender seedlings. Too much water at once can uproot them and they will die. Also, new seeds cannot take a hard flow of water. They need to be watered gently until they come up.

New Word:

Irrigation - To bring heavy flow of water by trenches into your garden, flood the garden, if you have a lot of water, or else run it down little ditches along each row.

1. **Flood Irrigation.** Flooding is best for big fields of plants such as corn, alfalfa, wheat and rice.

    a. The water will flow along the main trench to the field.

    b. You will open the wall of the field by digging away a small section.

    c. When the whole field is wet you will close the wall again.

2. Row Irrigation

    a. The water will flow along the main trench to the field.

    b. Open the wall so that the water runs down the lowest row.

    c. When the row is wet; block the water so that it will flow into the next row.

    d. As each row is wet, go to the next row until all the rows are wet.

The exact places for the trench must be carefully planned. The water has to flow downhill, but you want it to flow slowly so that it doesn't wash the plants. Make your trenches according to the slope of your land.

Always begin at the lowest end of your field.

Insect Control

1. The first rule for keeping insects out of your garden is to keep the garden clean. Remove all trash from your garden. Pick up all sticks, twigs, branches, leaves, old rags, pieces of wood or any other trash that may be lying around. Do not let the insects have a home were they can breed.

2. If you see insects in the garden you can pick them off with your fingers and kill them.

Spray Bugs Off

3. Wash the insects off with some water. Aphids and other sucking insects can be off the plant stems they have been feeding on. They will not return.

4. Make a poison by adding one cup of soap shavings or one cup of kerosene to a gallon of water. Then spray it on the plants.

5. Rotation of crops helps to control aphids and other insects.

6. Soil rich in humus helps control aphids.

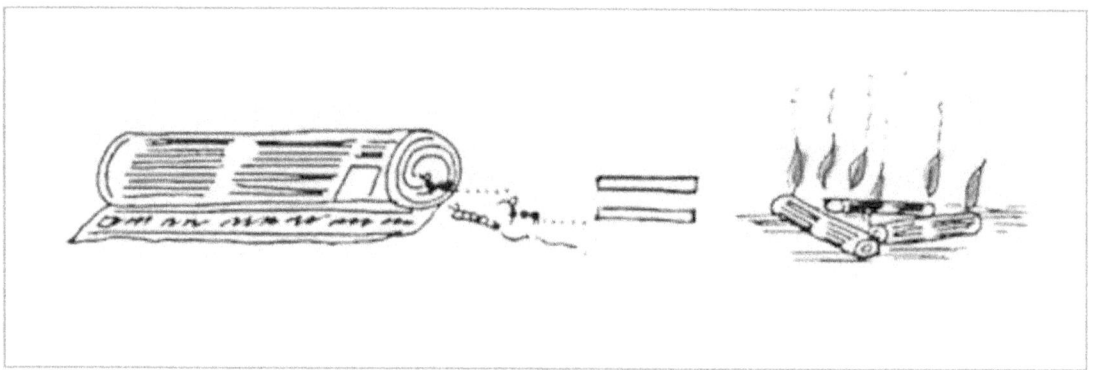

7. Place rolled up newspapers on the ground at night. During the night, crickets and other bugs will crawl into the newspaper. In the morning, you can burn the newspaper.

Good Bugs

8. Wasps, praying mantises, and lady bugs kill harmful insects. Don't kill wasps, praying mantises or lady bugs if you see them in your garden.

9. Plant sunflowers five feet outside your garden. The sunflowers will attract insects away from your garden.

10.Plant tomatoes close to the house, especially the kitchen to keep flies away.

11. Plant a border of peppermint around your garden to help keep ants away.

Good Garden Helpers

12.Birds, frogs, and lizards also kill insects. Let these animals live in your garden. But remember that birds can also eat your plants.

13.Plant a border of marigolds around the garden. The marigolds will keep insects away from the garden.

14.Sprinkle some ashes around the plants to keep cutworms and snails away

15.Plants that have just started to grow are often attacked by "cutworms". Cutworms eat plants right at the soil line. Place a small roll of paper around each plant to protect the stems from the cutworms.

16.Grasshoppers and other insects that chew leaves do not like the taste of onions. Mix some onion juice with a little bit of water and spray it on the plants to discourage chewing insects.

17.Smoke also drives insects away, but you have to be careful not to scorch or burn the plants.

Rodent Control

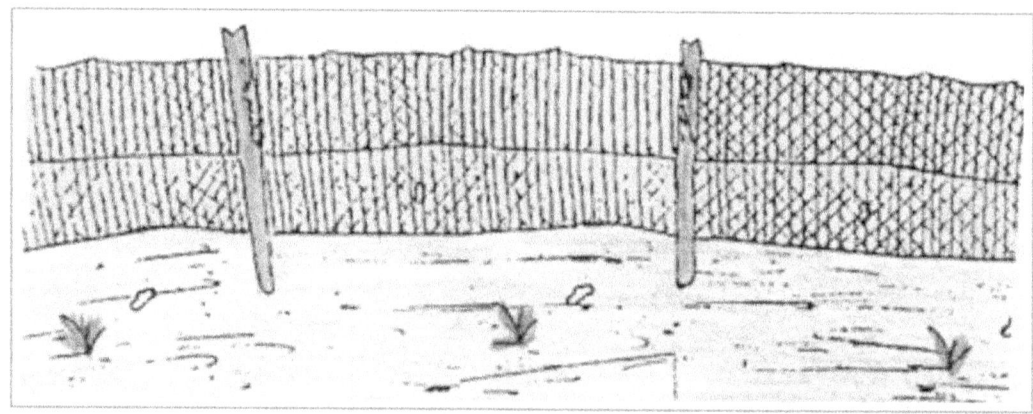

Rodents and some farm animals would like to get into the garden and eat plants. To keep them out you will have to build a fence. But some rodents can burrow under your fence. A 4 foot chicken wire fence will stop them if the fence is partly underground. Dig a trench 2 feet deep around your garden; put the wire into the trench - 2 feet below the ground and 2 feet above - fill the trench. A brick or cement block wall built the same way would also keep the rodents out.

Weeds

The only really effective way to get rid of weeds is to pull them out by hand. You must pull out the roots of the weeds. If you leave the roots, the weeds will grow again. The weeds will come out easier when you soak the ground before you pull them out. If the ground is too dry, you may pull out plants you do not want to, or leave some roots in the dry, hard ground. If you have a very big field, you can easily use a hoe to loosen the weeds, but be very careful not to cut the weeds up with the hoe and leave pieces that will take root and grow again.

Diseased Plants

Remove or treat all diseased plants as soon as you see the disease.

Harvesting

Replanting - After the crops have been harvested, you can replant right away. Make sure you add more compost or dry manure to the soil. Pull up the old plants and use them to make compost. Plant a small quantity of each crop at one time. Two weeks later, plant a small quantity of each crop. Do this as often as your planting season permits. This prevents waste of crops which cannot be used quickly enough.

Two weeks - Four weeks and Six weeks

Rotating Your Crops - Corn uses up a lot of nutrients in the soil. Beans, peanuts, and other legumes put nitrogen back in the soil. It is a good idea to plant beans or some other legume where the corn was the year before. Plant the corn in a different part of the garden for a year or 2 and then plant it again where the beans have been. This will keep your soil full of nutrients.

Tomatoes seem to like to be grown in the same place every year.

Harvesting Your Crops - As a rule it is best to pick vegetables while tender. Corn is ready to be picked in 65 - 90 days, depending on the type of corn.

Okra pods should be picked just before they are ripe. If they are not picked, the plant will not make new pods.

Beans are ready to be picked about 45 days after planting.

Carrots and onions can be used at almost any time after the first few weeks.

Pick cucumbers any time before they begin to turn yellow. You usually can tell if the crop is ready by the taste.

Seeds

Grow Your Own Seeds

Root vegetables are generally Biennials. This means it takes two years to make a seed. The first year it makes a root and stores food in it. The next year the root vegetables sends up a long stalk. It has all the stored food from a summer's growth in it's root to draw on. The stalk flowers. Where each flower was, seeds will form and can be collected when dry.

Fruited Plants make their seed inside the fruit. For example, melons tomatoes, green peppers, and eggplant. First comes the flower, and then the fruit. Let the seed - bearing fruit get fully ripe. Then scoop out the seeds and let them dry naturally - on a newspaper.

Tubers Are under the ground - for example potatoes. The plant sends up a stalk - flowers above the ground. This is not the seed the small knobs on the potatoes are used to grow new plants.

Flower Seeds - You can save seeds from almost any flower that has a flower head such as Marigold or Zinnia. Go out in the fall when the flowers are dry. Cut off and dry the flower heads. Store in a dry place for the winter. In the spring, break the head apart. The seeds are in the flower head.

Seeds saved from your garden have a

built in resistance to insects and diseases. These seeds also produce better tasting crops and are more nutritious. If you save the best seeds from your garden, you will not run out of seeds. You can save many seeds and share them with others. In fact, you might even sell some seeds for a profit.

Save the best seeds so you can plant them again and little by little make your crop better. Choose the best seeds from each harvest. Save the seeds from:

1. Strong, healthy plants.

2. Plants that grow fruit quickly.

3. Plants that don't get diseases.

4. Plants that have large or tasty fruit.

Remove sick or unhealthy plants from your garden before they make your healthy plants get sick.

Harvest the best seeds from fruits and vegetables that are well ripened, but not too ripe. Don't let them rot or blow away.

Collect the seeds on a dry day, after the dew is gone.

Harvesting and Cleaning Vegetables

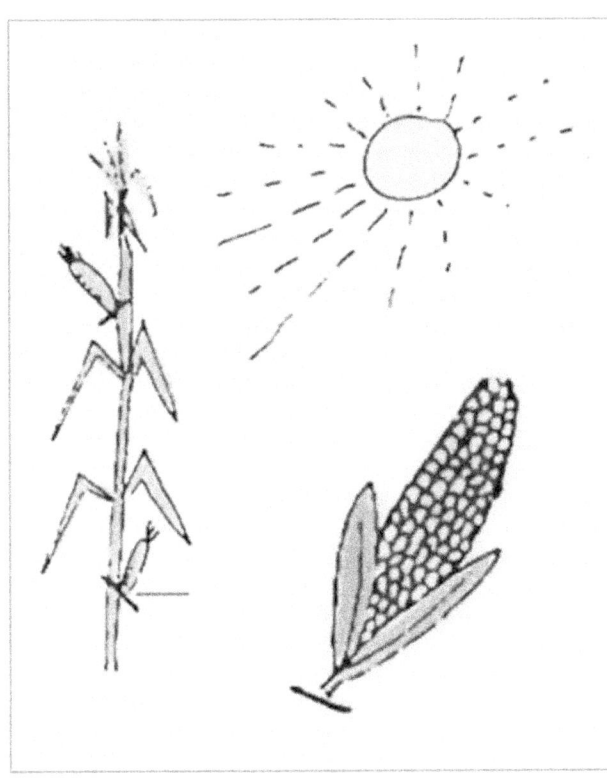

Corn: Pick the corn after the stems and leaves have turned brown. Take away the husks from the ears of corn. Put the corn in the sun to dry. When the corn is completely dry, put it away.

Okra: Remove stems when pods are dry, almost splitting. Remove seeds and dry before storing.

Squash: Pick squash, melons, and pumpkins when they are ready to eat. Wash the seeds and dry them as well.

Cucumbers: Should be picked when they are large and golden yellow. Wash the seeds and dry them.

Beans: Pull up the bean plants when the pods turn brown. Hang up the whole plant in a sheltered place to dry.

Peas: Pull up when the pods have turned brown. Hang up to dry.

Onions: Wait until you see black seeds on the dead flowers, then cut the stems. Dry the flowers and rub the seeds from the heads. Or save very small onions for replanting. Store small onions in a cool, dark, dry place.

Leeks: Same as onions.

Garlic: Same as onions and leeks.

Carrots: Same as onions. Or save carrot tops in cold place or even in cold water.

They will not save very long. You will have to plant them soon. Soak them in water

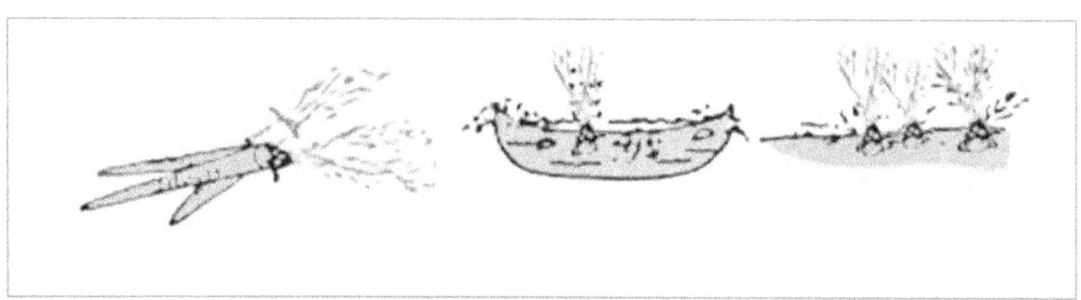

for a few days before you plant them.

Peppers: Remove the seeds and dry them.

Mustard: Cut the flower stems when the pods are very dry, but not wilted. Remove the seeds and dry them.

Radish: Same as mustard.

Chinese Cabbage: Same as mustard and radish.

Tomatoes: Pick tomatoes when they are red, but still firm. Place the seeds in a jar of water for two or three days. Good seeds will sink to the bottom. Pulp and bad seeds will rise to the top. Wash the seeds, then dry them on a screen or on newspaper.

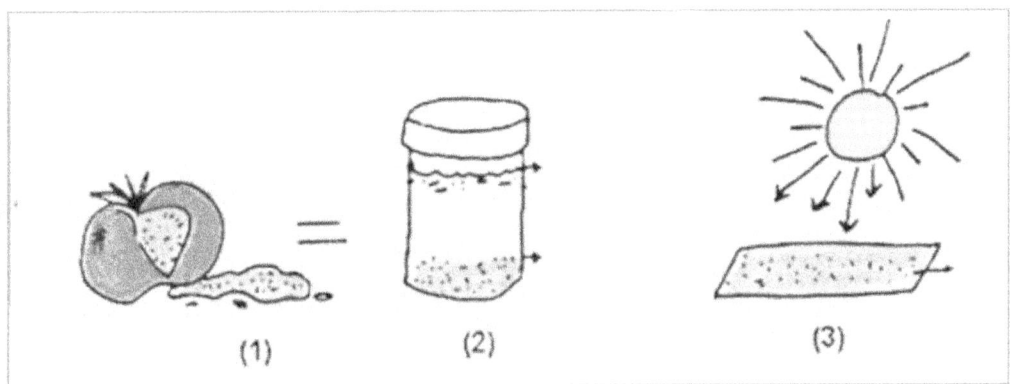

(1)   (2)   (3)

Eggplant: Same as tomatoes.

Potatoes: Pull up when plants are dry. Save the smallest potatoes for replanting. Keep some potatoes in a cool dry place and cut them into pieces, (each piece should have an eye (knob) when you are ready to plant them.

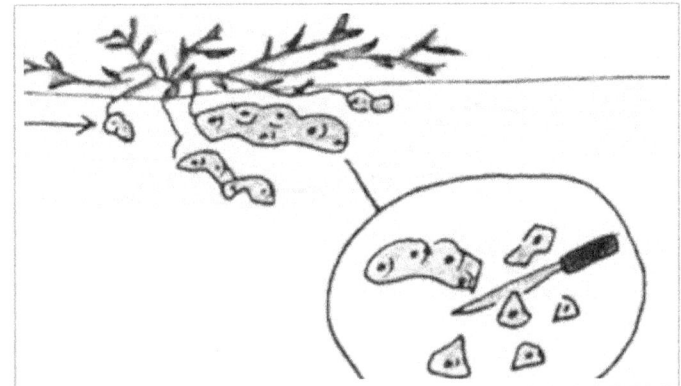

Storing Seeds

Seeds must be completely dry to prevent molding and being eaten by insects.

1. Test the seeds to see if they break with a "crack".

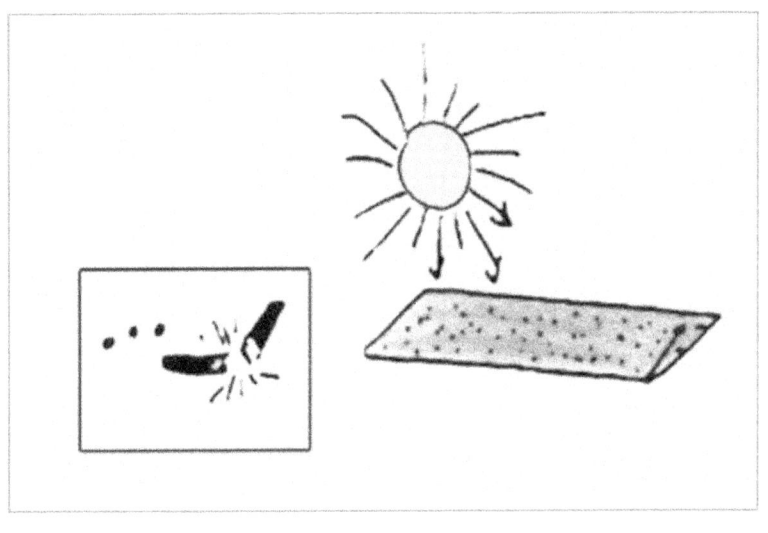

2. Mix some wood ash or lime with the seeds, if you can.

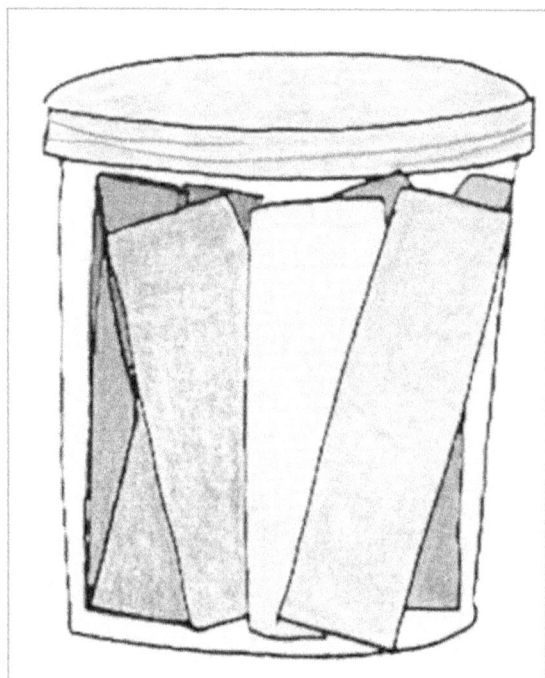

3. Save seeds in a jar that has a **tight lid**.

a. Several envelopes or plastic bags filled with seeds can be kept in the same jar. Put the date and name of seed on the envelope or jar. Each year use your oldest seeds first.

Or…

b. Place one type of seed in a small jar with **plastic seal on the lid.** Baby food jars are excellent for this. Remember to label each jar with the name and date.

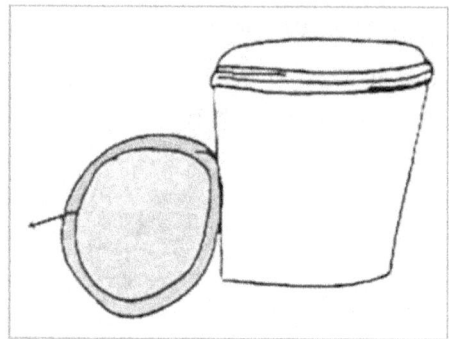

4. **Toasted rice** keeps moisture and water out of your dry seeds. If you live in a tropical country, it would be a good idea to mix some toasted rice in with the seeds.

How Long Will Seeds Last?

Most seeds will remain good for two to three years. Melon seeds last longer.

Since plants are living and growing things, they are very much like human beings. They need

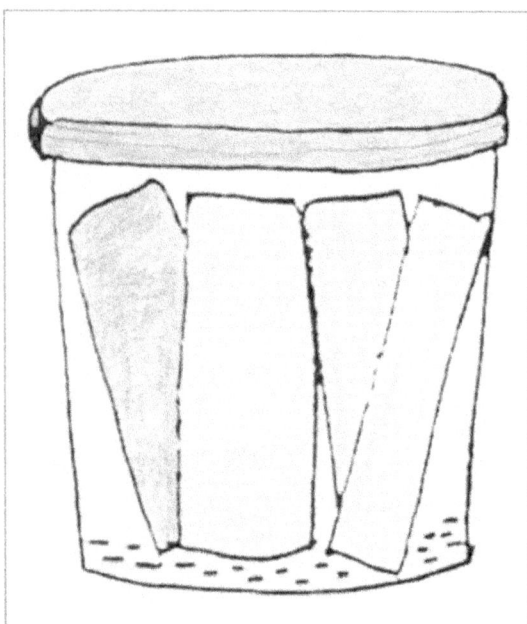

tender loving care. Jesus knew that when He said, "Consider the lilies of the field. They toil not, neither do they spin. Yet I say unto you that even Solomon in all his glory was not arrayed like one of these."

God cares for the plants. He gives them light and warmth from the sun. He waters them with the rain and gives them air to breathe.

He cares for people the same way.

And we should, too!

As God's children we should care for people by making sure that they have food, clothing, shelter, medicine and the Godly help that they need.

We need to share all the wonderful food that God has blessed us with, and thank Him for all He has given us.

Acknowledgments

Isaiah 58 Mobile Training Institute Spiritual Training Manual

© 2016 All Nations International All Rights Reserved

Isaiah 58 Mobile Training Institute books are available for use in training programs. to contact a representative please visit the Contact Us page at www.all-nations.org

Author: Irene Jensen

Artists: Irene Jensen, George Thomas, Jumi Sabbagh, Jennene Jeffrey, Sylvia Dawn, Virginia Russell, Teresa Skinner, Annella Whitehead

Let's Review

Close Together Planting

Fill in the blanks

1. _____ make more plants by forming _____.

2. Fruits are natural seed _____.

3. _____ make food.

4. Leaves draw in _____ _____ and lets out _____ and water vapor.

5. The _____ are plumbing.

6. _____ and _____ are taken from the soil through the _____

and are drawn up the stems to the _____.

7. The soil needs _____ and _____, and needs to be
"_____" and easy to dig and plant in.

8. You will want to be able to _____ your _____.

9. _____ needs a lot of water.

10. Most vegetables _____ best in direct _____.

11. Your plants need to have a minimum of _____ a day of sunshine.

12-14 List the three types of soil:

15. Good soil is a mixture of _____, _____, and _____ plants.

16. _____ is made up of tiny pieces of soil.

17. _____ is best for growing a garden.

18. _____ has larger particles, more rounded than flat.

19. If the _____ sticks together before it lands, there is too much _____ in the soil.

20. Another way to _____ the soil is to _____ it.

Score exercises 1 - 20

Underline the Correct Answer

21. (Close Together, Far Away) Planting is a way of gardening first used by (German, French) farmers centuries ago.

22. The farmers who started this kind of gardening had lots of (manicure, manure).

23. The way you plant your garden depends upon the kind of (weather, wither) you have in your (area, city).

24. If your climate is (dry, wet) you will want to make (hedges, rows). You will want to grow your vegetables (down, upside down) inside the trenches.

25. If your garden is in a (wet, dry) area, you will want to plant you garden on raised (beds, pillows).

26. The (roots, stems) of the plants will grow (deep, shallow) into the soil.

27. (Compost, Compose) is mixed into the soil helping the plants get (minerals, numerals) they need and providing nutrients at all levels.

28. If you have a lot of (waterfall, rainfall), plant seeds in rows along the (hedges, ridges).

29. It is (impossible, possible) to plant the seeds close together because the garden bed has been prepared by deep digging.

30. (Big, Small) vegetables like corn and squash are going to take (more, less) room no matter how hard you try.

Score exercises 21 - 30

Multiple Choice (Circle the letter for the correct answer):

31. Compost is:

a) rotted leaves, plants and dead insects.

b) Fresh leaves, plants, and live insects

c) Old leaves, plants, and insects

32. You should never use:

a) Smashed fruit or the skins, nor lettuce leaves in compost.

b) Citrus fruit or the peelings from them, nor cabbage leaves in compost.

c) Raw fruit or their peelings, nor corn cobs in compost.

33. Manure is:

a) Said to have a very bad smell

b) From a cow dung.

c) A very good compost.

34. The best manure for your garden is:

a) chicken manure

b) horse manure

c) rabbit manure

35. Humus in the soil helps beat down the aphid population and also helps so much when there is:

a) A drought

b) A storm

c) A hot weather

36. Compost also makes the fruits and vegetables taste:

a) sweet

b) sour

c) better

37. Find a place where water:

a) drains off quickly and make a compost pile there.

b) runs off quickly and make a heap pile there.

c) drains off slowly and make a garbage pile there.

38. Cover the whole pile with:

a) five inches of trash

b) two inches of vegetables

c) four inches of fresh manure

STOP! Score exercises 31 - 38

DO NOT look back at the Training Pac while completing the Review.

REVIEW # 1

Fill in the blanks

1. _____ make more plants by forming _____.

2. _____ make food.

3-5 List the three types of soil:

_____

_____

_____

6. Good soil is a mixture of _____, _____, and _____ plants.

7. Another way to _____ the soil is to _____ it.

Choose the best answer

8. (Close together, Far away) planting is a way of gardening first used by (German, French) farmers centuries ago.

9. If your climate is (dry, wet) you will want to make (hedges, rows). You will want to grow your vegetables (down, upside down) inside the trenches.

10. The (roots, stems) of the plants will grow (deep, shallow) into the soil.

11. If you have a lot of (waterfall, rainfall), plant seeds in rows along the (hedges, ridges).

12. (Big, Small) vegetables like corn and squash are going to take (more, less) room no matter how hard you try.

13. The way you plant your garden depends upon the kind of (weather, wither) you have in your (area, city).

Multiple Choice (Circle the letter for the correct answer):

14. You should never use:

a) Smashed fruit or the skins, nor lettuce leaves in compost.

b) Citrus fruit or the peelings from them, nor cabbage leaves in compost.

c) Raw fruit or their peelings, nor corn cobs in compost.

15. The best manure for your garden is:

a) chicken manure

b) horse manure

c) rabbit manure

STOP! Score exercises 1 - 15

True or False (Write "T" if the answer is True and "F" if the answer is False)

1._____ Planning your garden has a lot to do with where you live.

2._____ When you plant your garden make sure your plants will get enough sun but not too much sun. Too much sun will burn your plants.

3._____ If you live where the sun is very hot, you need to plant the smaller plants toward where the sun rises, then the short plants will help shade the taller plants.

4._____ Some plants like corn need the hot sun.

5._____ Remember to keep the garden bed soaked at all times while the seeds are beginning to grow.

6._____ When the plants are three weeks old, you will only need to water them every three days, but you will not water them lightly -- instead you will water them until the garden bed is thoroughly soaked.

7._____ Rain is God's way of watering a garden.

8._____ If your garden is in a wet area, you will want to plant your garden on "flat beds". This way, the plants are below the water and will not be washed away.

9._____ Seed started in flats cannot be completely protected from the weather, insects, and weeds until they are strong enough to be transplanted into the garden.

10. _____ Transplant early in the afternoon so that the plants will get some sun but not too much.

11. _____ When you have transplanted all the plants, water the bed gently but thoroughly.

12. _____ Keep the garden moist at all times when the plants are first beginning to grow.

13. _____ Mulching keeps your plants very wet and helps keep grass away.

14. _____ Flowers can survive longer without water than vegetables, so can trees and bushes.

15. _____ A good way to decide where your garden should be and how large it should be is to let the sun decide the size.

Score exercises 1 - 15

Fill in the blanks

16. Another way to water is to carry the water in _____ and pour it beside the plants.

17. Remember to be very _____ on young tender _____.

18. _____ is best for big fields of plants such as corn, alfalfa, wheat and rice.

19. The water has to flow _____, but you want it to flow _____ so that it doesn't wash the plants.

20. The first rule for keeping _____ out of your garden is to keep the garden _____.

21. _____ and other sucking insects can be _____ off the plant stems they have been feeding on.

22. _____ of crops helps to _____ aphids and other insects.

23. Wasps, _____ _____, and lady bugs _____ harmful insects.

24. _____ and some _____ animals would like to get into the garden and eat plants.

25. The only really _____ way to get rid of weeds is to pull them out by _____.

Score exercises 16 - 25

26-36 Match the correct Answer (Match the statement on the left to the letter on the right.)

Statement	Match to Correct Letter
26. After the crops have been harvested	a) by the taste.
27. It is a good idea to plant beans or some other legume	b) you can replant right away.
28. Tomatoes seem to like to be grown	c) where the corn was the year before.
29. Beans are ready to be picked about	d) 45 days after planting.
30. You can tell if the crop is ready	e) in the same place every year.
31. Root vegetables are generally Biennials.	f) built in resistance to insects
32. The small knobs on the potatoes	g) are used to grow new plants.
33. Seeds saved from your garden have a	h) This means it takes two years to make a seed.
34. Remove sick or unhealthy plants from your garden	i) the pods turn brown.
35. Pull up the bean plants when	j) to prevent molding and being eaten by insects.
36. Seeds must be completely dry	k) before they make your healthy plants get sick.

STOP! Score exercises 31 - 36

DO NOT look back at the Training Pac while completing the Review Close Together Planting WORKBOOK

REVIEW # 2

True or False (Write "T" if the answer is True and "F" if the answer is False)

1._____ Planning your garden has a lot to do with where you live.

2._____ If you live where the sun is very hot, you need to plant the smaller plants toward where the sun rises, then the short plants will help shade the taller plants.

3._____ Rain is God's way of watering a garden.

4._____ A good way to decide where your garden should be and how large it should be is to let the sun decide the size.

Fill in the blanks

5. Remember to be very _____ on young tender _____.

6. The first rule for keeping _____ out of your garden is to keep the garden _____.

7. The only really _____ way to get rid of weeds is to pull them out by _____.

8. _____ seem to like to be grown in the same place every year.

9. Root vegetables are generally _____. This means it takes _____ years to make a seed.

10. _____ must be completely dry to prevent _____ and being eaten by insects.

Score exercises 1 - 10

STOP! You must now **prepare yourself for the PRE-TEST**. In preparation, you may want to follow one or more of these suggestions:

1. Rewrite every incorrect exercise in the Reviews.

2. Reread each section of the Training Pac.

3. Relearn each section you still do not completely understand.

Pre-Test

Fill in the blanks

(5 points each question)

1. _____ make more plants by forming _____.

2. _____ make food.

3. Good soil is a mixture of _____, _____, and _____ plants.

4. _____ and _____ are taken from the soil through the _____ and are drawn up the stems to the _____.

5. Most vegetables _____ best in direct _____.

Underline the best answer

6. (Close together, Far away) planting is a way of gardening first used by (German, French) farmers centuries ago.

7. The way you plant your garden depends upon the kind of (weather, wither) you have in your (area, city).

8. If your garden is in a (wet, dry) area, you will want to plant you garden on raised (beds, pillows).

9. (Compost, Compose) is mixed into the soil helping the plants get (minerals, numerals) they need and providing nutrients at all levels.

10. (Big, Small) vegetables like corn and squash are going to take (more, less) room no matter how hard you try.

Multiple Choice

11. Compost is:

a) rotted leaves, plants and dead insects.

b) Fresh leaves, plants, and live insects

c) Old leaves, plants, and insects

12. MANURE is:

a) Said to have a very bad smell

b) From a cow dung.

c) A very good compost.

13. Humus in the soil helps beat down the aphid population and also helps so much when there is:

a) A drought

b) A storm

c) A hot weather

True or False (Write "T" if the answer is true and write "F" if the answer is false)

14. _____ Planning your garden has a lot to do with where you live.

15. _____ If you live where the sun is very hot, you need to plant the smaller plants toward where the sun rises, then the short plants will help shade the taller plants.

16. _____ Remember to keep the garden bed soaked at all times while the seeds are beginning to grow.

17-20 Match the correct Answer (Match the statement on the left to the letter on the right.)

Statement	Match to Correct Letter
17. The first rule for keeping insects out of your garden	a) This means it takes 2 years to make a seed
18. Rotation of crops helps	b) to prevent molding and being eaten by insects.
19. Root vegetables are generally Biennials	c) control aphids and other insects
20. Seeds must be completely dry	d) is to keep the garden clean

Let's Review Key

Section I

1. Flowers, seeds

2. holders

3. Leaves

4. carbon dioxide, oxygen

5. stems

6. Nutrients, water, roots, leaves

7. air, water. loose

8. water, garden

9. Vegetables

10. grow, sunlight

11. 6 hours

12. clay

13. sand

14. loam

15. clay, sand, rotted

16. Clay

17. Loam

18. Sand

19. ball, clay, soil

20. test, water

Section II

21. Close together, gardening, French

22. manure

23. weather, area

24. dry, rows, down

25. wet, beds

26. roots, deep

27. Compost

28. rainfall, ridges

29. possible

30. Big, more

Section III

31. A

32. B

33. C

34. B

35. A

36. C

37. A

38. C

Review #1

1. Flowers, seeds
2. Leaves
3. Clay
4. Sand
5. Loam
6. clay, sand rotted
7. test, water
8. Close together, French
9. dry, rows, down
10. roots, deep
11. rainfall, ridges
12. Big, more
13. weather, area
14. B
15. B

Section IV

1. T
2. T
3. F
4. T
5. F
6. T
7. T
8. F

9. F
10. F
11. T
12. T
13. F
14. T
15. F

Section V

16. bucket
17. gentle, seedlings
18. Flooding
19. downhill, slowly
20. insects, clean
21. Aphids, washed
22. Rotation, control
23. praying mantises, kill
24. Rodents, farm
25. effective, hand

Section VI

26. B
27. C
28. E
29. D
30. A

Section VII

31. H
32. G
33. F
34. K
35. I
36. J

Review #2

1. T
2. F
3. T
4. F
5. gentle, seedlings
6. insects, clean
7. effective, hand
8. Tomatoes
9. Biennials, 2
10. Seeds, molding

6. Close together, French
7. weather, area
8. wet, beds
9. Compost, minerals
10. Big, more
11. A
12. C
13. A
14. T
15. F
16. F
17. D
18. C
19. A
20. B

Pre-Test

1. Flowers, seeds
2. Leaves
3. clay, sand, rotted
4. Nutrients, water, roots, leaves
5. grow, sunlight

Final Test

Word List		
Weather	Main Roots	Close Together
Moist	Rainfall	Rotted
Horse Manure	Hot	Mulching
Hand	Transplanted	Rows
Biennials	Clean	Sunlight
Rotation	Compost	Vegetables
Flowers	Molding	

Fill in the blanks with the words from the word list

1. _____ make more plants by forming seeds.

2. _____ anchor and hold the plant upright against wind and weather.

3. Most vegetables grow best in direct _____.

4. _____ planting is a way of gardening first used by the French farmers centuries ago.

5. The way you plant your garden depends upon the kind of _____ you have in your area.

6. _____ is mixed into the soil helping the plants get minerals they need and providing nutrients at all levels.

7. If you have a lot of _____, plant seeds in rows along the ridges.

8. Compost is _____ leaves, plants and dead insects.

9. The best manure for your garden is _____.

10. If you live where the sun is very _____, you need to plant the taller plants toward where the sun rises, then the tall plants will help shade the shorter plants.

11. Remember to keep the garden bed _____ at all times while the seeds are beginning to grow.

12. Plant the seeds in _____.

13. When the second set of leaves begin to appear, the plants can be _____.

14. _____ keeps your plants moist and helps keep weeds away.

15. _____ need a lot of water.

16. The first rule for keeping insects out of your garden is to keep the garden _____.

17. _____ of crops helps to control aphids and other insects.

18. The only really effective way to get rid of weeds is to pull them out by _____.

19. Root vegetables are generally _____. This means it takes two years to make a seed.

20. Seeds must be completely dry to prevent _____ and being eaten by insects.

Final Test Key

Final Test

1. Flowers

2. Main Roots

3. Sunlight

4. Close Together

5. Weather

6. Compost

7. Rainfall

8. Rotted

9. Horse Manure

10. Hot

11. Moist

12. Rows

13. Transplanted

14. Mulching

15. Vegetables

16. Clean

17. Rotation

18. Hand

19. Biennials

20. Molding

CHAPTER 3

Raising Goats

Goats can be profitable if you take care of them.

There is a big difference in people that just watch goats and people that raise goats. This book is for people that raise goats. It is them that want to make improvements so they can make more profit from raising goats.

It is my hope that this book can benefit those of you who raise goats and you can make progress. I hope that after you study and apply what is in this book that you will help other people to do a better job of raising goats. Maureen E. Birmingham DVM

Preventing Sickness and General Care

To prevent sickness and keep the goat in good shape: Keep the goats in a place that is clean, dry, and has shelter. Goats do not like mud or rain.

Give the goats clean fresh water to drink. Goats won't drink dirty water. Goats need water to grow faster and give more milk. Give the goats forage that is green and young along with what other feed you can find. Don't change the diet fast. This may make them sick. If you must change the feed, do it slowly.

Give the goats wormer every three months.

Here is a list of wormers:

Fenbendazole	(Panacur, Safeguard)	10-15 mg/kg
Mebendazole	(Telmin, Vermox)	15-20 mg/kg
Levamisol	(Ripercol, Levasol)	7.5 mg/kg
Tetrasole	(Tramisol)	15 mg/kg
Ivennect'm	(Ivomec)	1 cc/ 110 pounds
Albendazole	(Valbazen)	1 cc/30 pounds

Don't drink the milk for 4 days after giving the wormer.

• If there is a problem with tetanus is the area, give the goats a tetanus vaccination every year.

• You should vaccinate the doe 3-6 weeks prior to kidding.

• Vaccinate the kids when they are 1 month old and then again at two months.

• When vaccinating a goat for the first time, give a second dose one month later.

• Trim the hooves often to prevent infection and deformation.

• Check your goats every day.

Goats don't like mud or rain. They can get depressed.

Goats need a place that is well drained.

They should have a shed for protectionfrom the sun and rain.

PREVENTING PROBLEMS WITH WORMS

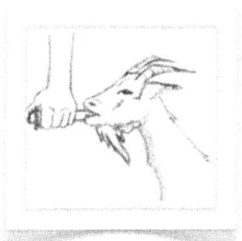

Give dewormer every three months

When you feed forage tie it up or put it in a feeder so that it can stay clean and off of the ground.

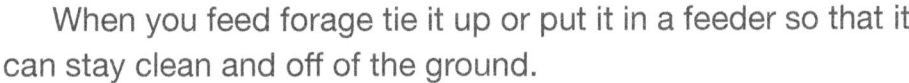

Stake the goats in clean areas that have not been contaminated by other goats change the place every day.

How To Trim Goat Hooves

When the hooves get like this, they need to be trimmed so they don't get infected or deformed

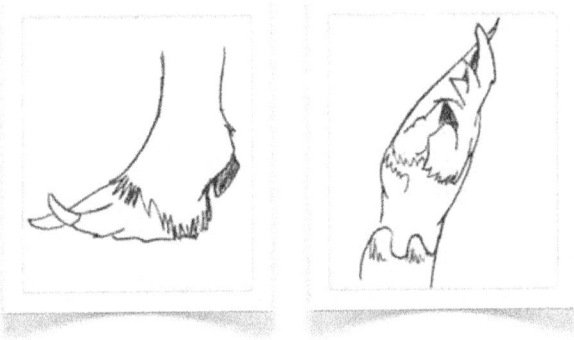

Dig dirt out from toes.

Trim away all loose, excess nail. Trim parallel to hoof hairline.

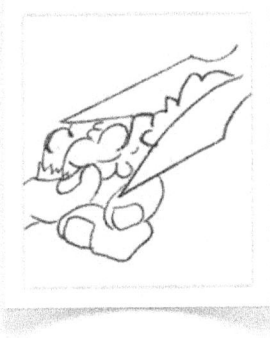

Pare heels to same level as toes.

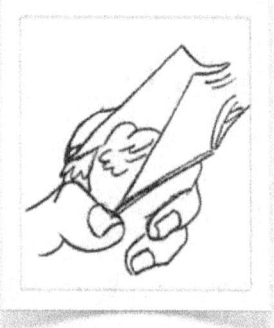

Snip away the little flap that grows between the toes.

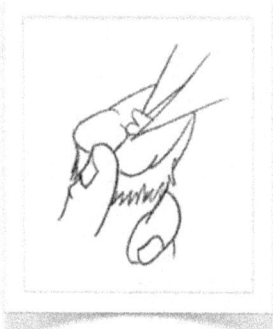

Finish the trim by paring the soft heel tissue til hoof surface is smooth and flat.

Finished.

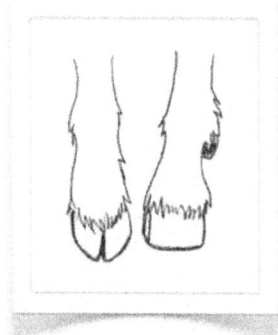

How To Hold A Goat While Trimming Its Hooves

Breeding And Delivery

Breeding

Goats come into heat between 5 and 10 months of age.

Most goats come into heat in the months of April May, June and in August, September, and October. There are some that will come into heat at any time.

It helps to feed the doe extra for three week prior to breeding to improve chances for more kids.

Goats come into heat every 21 days and stay in heat for I to 3 days. The doe will seem agitated, bleat a lot, urinate frequently, and will be interested in the buck.

Breed the doe with the buck twice while it is in heat, or turn the buck in with it.

The doe takes 5 months to kid after breeding. If you breed it 20 March, it will kid about 20 August. Give the doe more feed for six week prior to kidding.

Delivery

If every thing goes well, the doe will kid in one to two hours time after it starts labor. If the doe has not kidded one hour after the water breaks, raise its rear end to help the kid change position inside. This may help the doe kid on her own.

If the doe still has problems, wash your hand and arms. Then wash the does vulva and gently enter your hand and try to pull the kid. After helping the doe, give her antibiotics (I cc penicillin/30 pounds daily for three days).

Most does will shed the placenta within 4 hours time after kidding. If she has not shed the placenta after four hours, give her 1/2cc oxytocin every 2 hours. Also give her procaine penicillin daily for 3 to 5 days. If' you have benzathine penicillin every two days (I cc/30 pounds) or I cc LA 200 per 20 pounds every 3 days after delivery.

Within 12 hours Before Delivery The Doe Will Start To Show Signs That Delivery Is Close

The doe stays by herself. She is nervous. She paws the ground.

There is a depression. There is a thick mucous discharge from the vulva. The udder is full. Colostrum may wax at the end of the teat The stomach drops.

She lays down often. She looks at her side. She is uneasy. She bleats a lot

Normal Delivery

When you see signs of labor put the doe in a place by herself that is clean and

has straw.

Her water breaks. The kid should appear within an hour after the water breaks.

The kid should be born within two hours time after the water

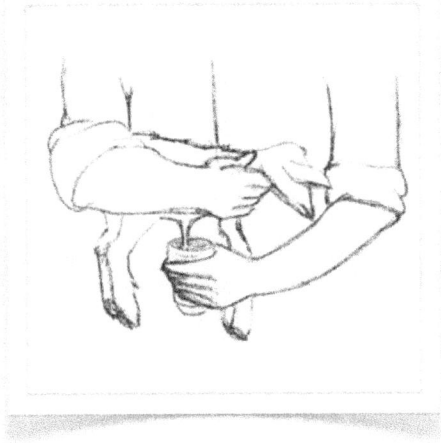

Dip the navel chord in disinfectant. (Tincture of iodine).

Let the kids nurse as soon as possible so they can get colostrum. Colostrum is the

first milk from the mother. The colostrum will protect the kid from many diseases.

Delivery Postions And Dystocia

Breeding

Position of the kid: The head is between the front legs facing forward. The front legs come out first.

Rear legs come out first

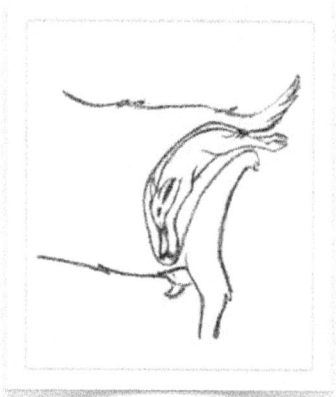

Both kids are in normal position in the uterus

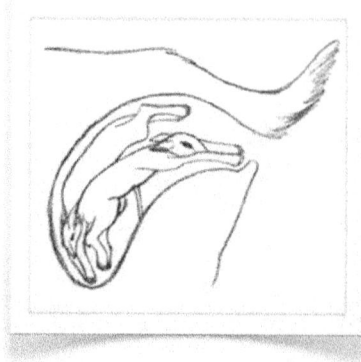

.Causes of Dystocia

Abnormal Positions at Birth

Push the kid forward and pull the rear legs back so the kid can come out.

Rear of the kid coming first.

The head turned back. Push the kid forward and turn the head so that it is positioned between the leg and pull the kid.

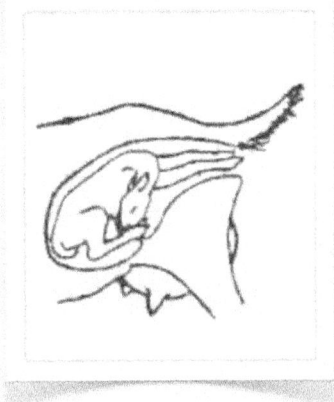

If only one leg appears and the other is turned backward the kid may not be able to deliver.

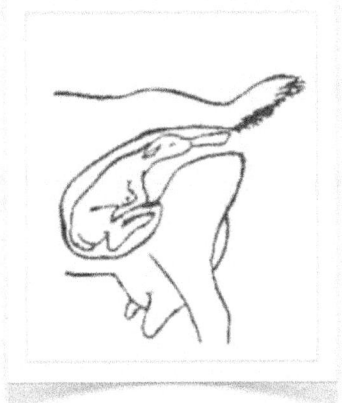 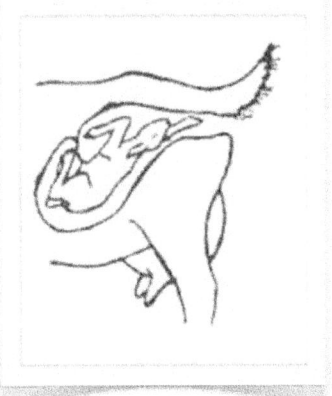

Helping a Goat with Dystocia

Cut the fingernails very short

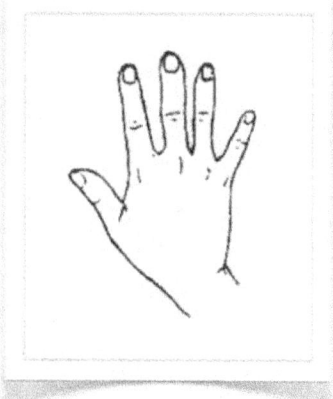

Wash the hands and arms very well with soap and water.

Wash the rear of the doe with soap and water.

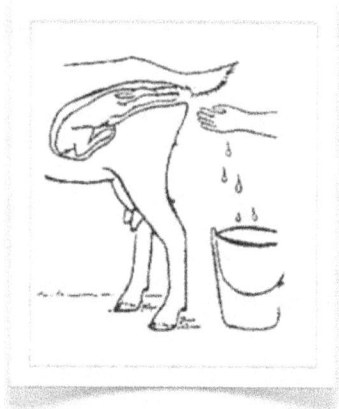

1. Push your hand inside very gently.

2. Move the kid inside the uterus so it can be delivered.

3. Pull the kid by the front legs. The front legs should come out before the head.

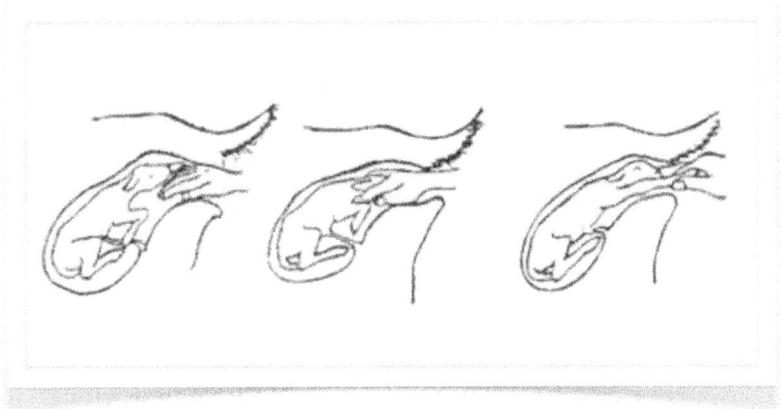

When the doe finishes delivery:

Give the doe antibiotics for 5 to 6 days.

Example:

Give 3cc of penicillin G every day for 5 days or 3cc LA 200 every 3rd day for 6 days.

After Delivery

The doe should shed her placenta within 4 hours time after delivery.

If she doesn't, give her 1/2 cc oxytocin every 2 hours the first day. Also give her antibiotics.

The doe needs **more** water to produce milk. If she doesn't get enough water she won't have milk for her babies. Give fresh, clean water 3 times per day.

The doe also needs more feed of good quality to produce plenty of milk.

Care Of The Kids

When the kid is first born, dip the umbilical cord in a disinfectant such as iodine or alcohol. This can help prevent bacteria from entering the cord.

Help the kid nurse as soon as possible. It is very important for the kid to get colostrum as soon as possible. This will help protect it from infections and will help it pass the meconium.

De-bud the goats if they will be in a pen with others so they don't damage others with their horns. This should be done before 2 months. If you wait too long, it is much more difficult. If the goats will be in a large pasture, the horns may be good protection against dogs and wild animals.

It is best to castrate the male goats before they are 3 weeks old.

If there is a problem with tetanus in the area, the kids should be vaccinated against tetanus at one month of age and again at two months.

Dip the umbilical chord in disinfectant.

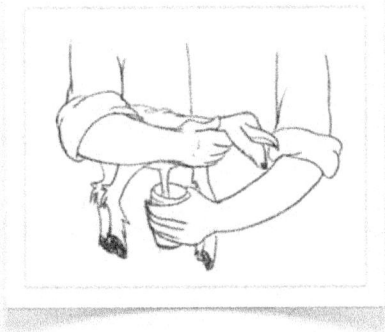

Let the kid nurse quickly to get colostrum.

De-Budding

The kids should be de-budded if they are going to be kept in a pen, so they will not hurt the other goats. This should be done as soon as you see the horn buds.

Heat up a metal pipe 1/2 inch in diameter.

After it is very hot, place it on the small buds for 15 seconds. You need someone to hold the kid very still to keep it from being injured.

If it is burnt properly you will see a mark around the horn bud.

Attention: be careful not to burn the bud too much and damage the brain.

Castrating A Goat

There are two ways to castrate a goat. You can crush the cord or you can cut the scrotum and remove the testicles. You can crush the cord with two pieces of wood or an instrument called a burdizzo.

1.Location to crush the cord. Clean the scrotum with soap and water.

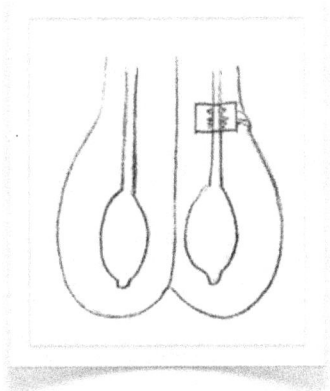

2.Cut it with a clean sharp blade. Pull the testicles and cut the cord. If the goat is large, you can tie off the blood vessels before cutting. Spray with insecticide and give antibiotics for three to four days for infection and tetanus.

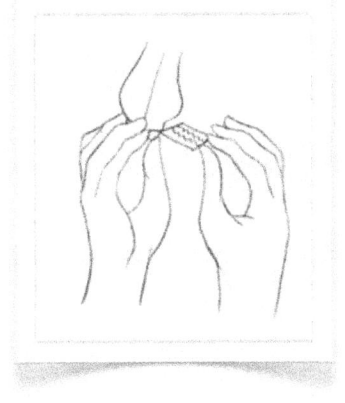 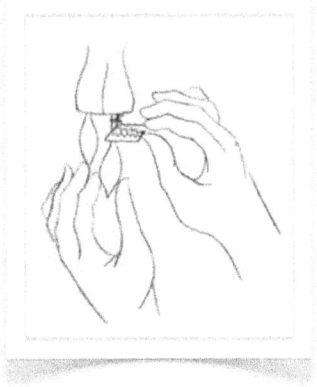

Choosing A Goat To Buy

Choose A Goat That Is:

Not too fat

Not too thin

Do Not Choose A Goat That Has:

1. Weak Back

2. Rump that is too sloped and weak.

3. Sagging Udder

4. Weak Legs and Fetlock

5. Flat Stomach

6. Straight Shoulders

7. Too Old: Check teeth. Narrow nose and mouth.

Choose A Goat That:

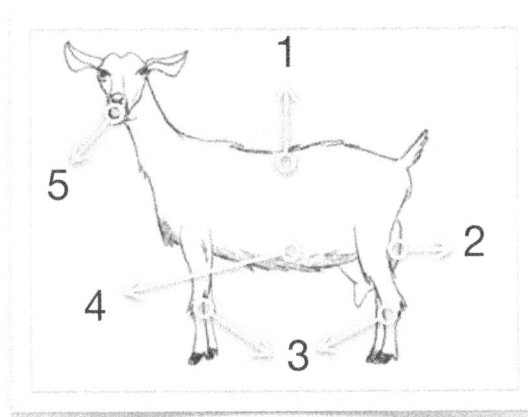

1. A Strong Back

2. Good Udder

3. Strong Legs

4. A stomach that has a large capacity

5. Good Teeth

Good: Legs that are straight

Bad Legs are not straight

Good

Bad

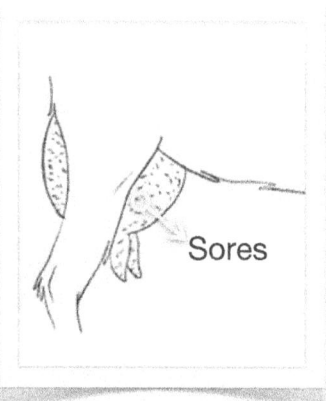

Sores

This udder is weak and the teats are too big.

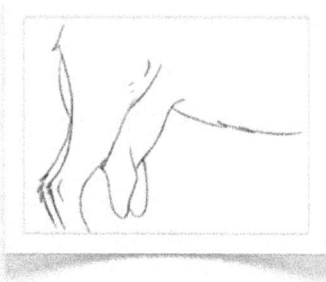

How To Choose A Buck To Buy:

A Good Buck Has:

1. Clear Eyes

2. Large Head with a Long Mandible

3. Large Mount and Nose

4. Heavy Muscle

5. Large and Strong Body

6. Strong Back with Large Rump

7. Large Testicles that are the Same Size

8. Good Teeth

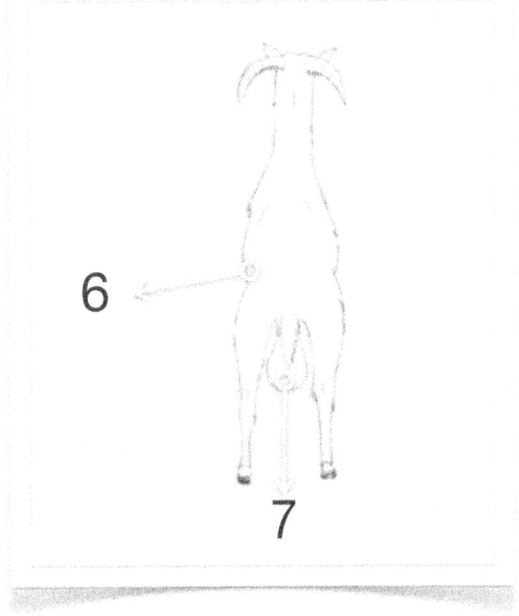

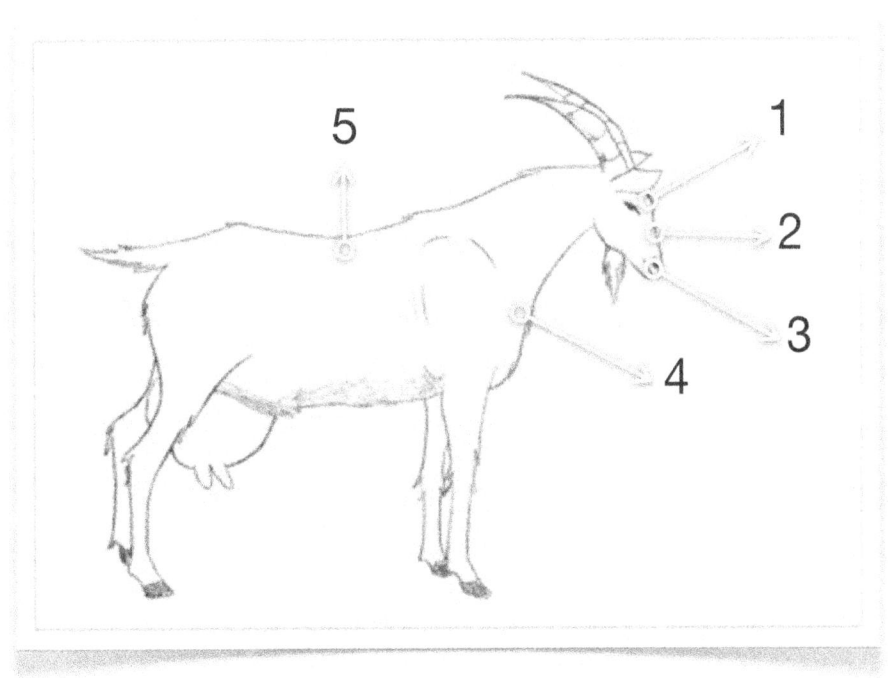

Aging A Goat

To Age A Goat Look At The Lower Incisors

(Front Teeth)

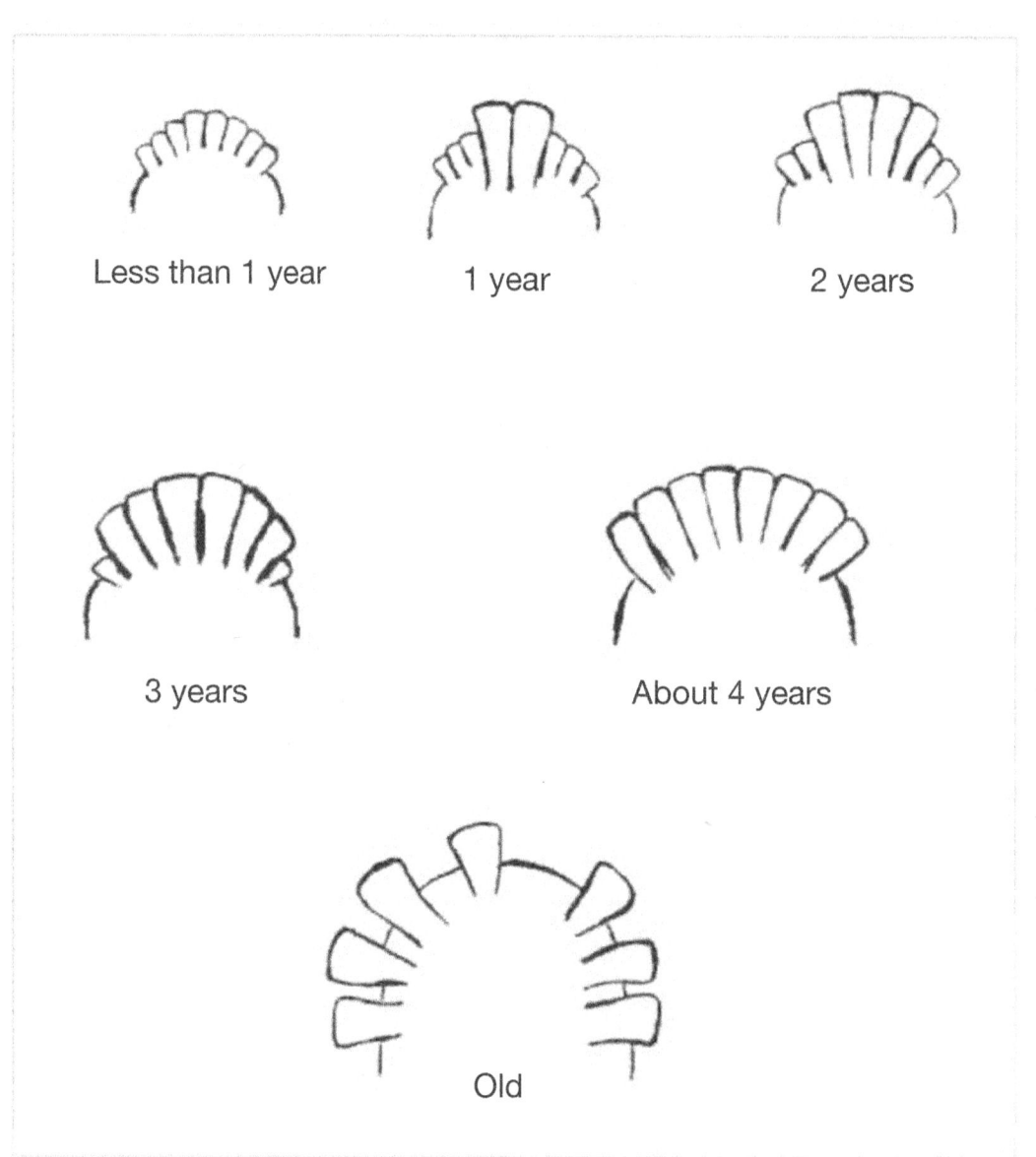

Less than 1 year   1 year   2 years

3 years   About 4 years

Old

Weighing A Goat

Weighing A Goat Without A Scale

Measure the size of the goat with a measuring tape behind the front legs. Tighten the tape before you take the measurement.

Now look at the table to get the weight:

CENTIMETERS	INCHES	POUDS
25	10	4
28	11	5
30	12	6
33	13	8
36	14	10
38	15	12
41	16	15
43	17	19
46	18	23
48	19	27
51	20	31
53	21	35
56	22	39
58	23	45
61	24	50
63	25	57
66	26	63
68	28	69
71	28	75
73	29	81
76	30	87
79	31	93

CENTIMETERS	INCHES	POUDS
81	32	101
84	33	110
86	34	120
89	35	130
91	36	140
94	37	150
96	38	160
99	39	170
102	40	180
104	41	190
107	42	200

Goat Pens, Forage and Hay

When you keep a goat in a pen:

- •YOU can control the breeding.

- •YOU can watch the goats better.

- •YOU must find feed for the goats

- •It is easier to give clean water and good hay.

But to get the benefits of a pen:

- •YOU must give them fresh clean water everyday.

- •YOU must plant forage for the goats.

- •YOU must conserve forage for the dry season. This is called hay.

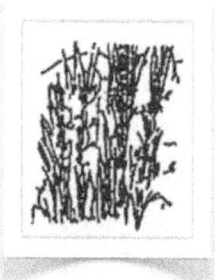

•You must give worming medicine at least every three months

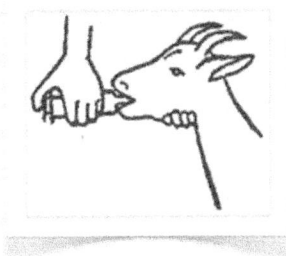

It Is Good to Give The Goats Salt

Put the salt in a gourd. Hang the gourd at a height the goat can reach.

The salt will soak through the gourd. The goat will lick the salt from the gourd.

When you keep your goats in a pen, you must cut forage and bring it to them.

It is better to put the hay in a feeder or tie it up so the goat will have clean feed that isn't contaminated with worm eggs. Goats will not eat dirty food.

 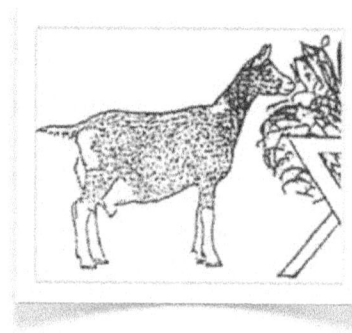

Making Good Hay

It is best for the forage to be green and young to make the best hay. When you make hay, you will conserve it for the dry season. Hay will keep the goats in dry season when there isn't good grass for the goats.

Spread the forage and turn it daily so that sun can dry it. Turn the hay gently so that the leaves do not fall off.

How You Can Lose Hay

• Don't make hay from old, dry forage. Young green forage makes the best hay and has more nutrition. To make the hay tastier, put a small amount of salt between each layer.

• Rain will make the hay spoil.

• It will lose it's leaves if it is too dry.

• If it is stacked wet, it will get hot and spoil.

• Put the hay inside or under a shelter so the hay won't get wet, because if it gets wet it will spoil.

Fencing For A Goat

You may use a yoke to help keep the goats from getting out.

Use living trees for a post.

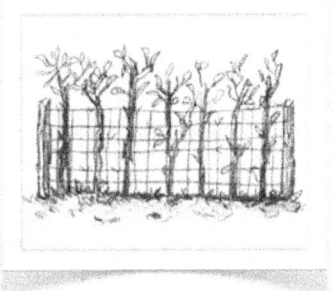

Use cactus or prickly vegetation

Don't Let Goats Run Loose

Goats running loose will kill small trees and destroy gardens. Keep goats on a rope or in a pen

Signs Of Sickness Or Disease

A goat that is sick:

·May not want to drink or eat.

·May a large head.

·May have diarrhea.

·May be anemic. Check the conjunctiva around its eyes. If it is white, the goat is anemic.

·May have a fever. A temperature above 104 F or 40 C.

·May be lame.

·May show signs of pain.

·The goat may move slowly or may kneel on its knees because the legs hurt.

·He may kick at its stomach or stand with its back humped up because the abdomen hurts.

·Breathes fast or with difficulty.

·Stays by itself or lays down a lot.

·Is bloated.

·Blind.

·Nervous.

·Has a vaginal discharge.

·Rough dry hair.

·Swellings on the body.

INTERNAL PARASITES

Stomach and intestinal worms: When a goat is depressed, weak, is anemic, has a swollen head and neck, has diarrhea, and has rough, dry hair she is likely to have internal parasites. These worms suck the blood from the goat and make them anemic and weak. When a goat is anemic, their mucous membranes will be white or pale.

Treatment: Give the goat worming medicine every month for three months and then once every three months after that. You should alternate types of wormers each time so that the parasites do not build a resistance against one type of wormer. Dose is on page 6.

Lung worms: There are two types of lung worms, muellerius and dictyocaulus.

The goats will get depressed and weak and may also cough. One must use a microscope to know for sure that the goat has lung worms.

Treatment: Fenbendazole and Ivermec are good for muellerius. Levamisole and Ivermec will work against dictyocaulus. See page 6 for the dosage.

Liver flukes: Fasciola: A goat with liver flukes may be depressed, thin, and anemic. You may suspect liver flukes if goats graze where there is standing water with snails, because snails carry the parasite.

Treatment: Abendazole is given at 1 cc/30 pounds for three days.

Note: It is much better to prevent internal parasites by grazing on clean pastures and rotating pastures. Worm a newly purchased animals before putting it with the others.

HOW A GOAT CAN GET COCCIDIOSIS

This goat has intestinal worms.

When it defecates, it contaminates the grass with worm eggs.

This goat comes and eats where the other goat has defecated.

While it is eating, it picks up the worm eggs.

The worm eggs develop in the stomach.

Now this goat has worms also.

Coccidiosis:

You will see this most often in kids under 4 months of age. Coccidiosis attacks the small intestine and makes the goat weak with diarrhea and can kill them. The diarrhea may have blood in it. Microscopic confirmation of the feces is the best way to know if the goat has Coccidiosis.

Treatment:

Sulfa: Give 25 mg/ pound per day orally for three days. There are different kinds of sulfa...Sulfamethamethazine, sulfaguanidine, sulfadimethoxine, etc.

Amprolium: Give it orally at 5 mg/ pound once daily for 5 days. B-Complex or B-12 vitamins may need to be given to help build the blood.

EXTERNAL PARASITES

Lice: This is a small insect you might find in the hair of a goat. They are very small and hard to see, but they make the hair rough and the animal scratches a lot.

They feed on the goat's blood and can make them anemic. Like internal parasites, this is a common problem.

Treatment: There are insecticide sprays, powders, or injections for lice.

Ivomec: This is an injectable and is dosed at 1 cc/ 110 pounds. Give the shot subcutaneously. This works very good and one treatment is usually enough. All other goats in the herd should be treated.

Coral (coumaphos): This is a powder to dust the animals.

Malathion 57% (liquid): Put two tablespoons in a gallon of water and bathe or spray the animal.

Malathion 24% (powder): Mix seven tablespoons in a gallon of water for bathing or spraying.

Lindane 12.5%: Mix one tablespoon in a gallon of water for bathing or spraying.

Asuntol (coumaphos): Mix one teaspoon in a gallon of water for bathing or spraying.

When using powder or spray, treat the animal every two weeks for three times.

Attention: These insecticides are poisons and should be handled carefully. Wear rubber gloves and a mask. Wash your hands with soap and water after working with these. Mark the bottle well and keep it out of reach of children.

Ticks: You can see these small anthropods on the animal. They also transmit anaplasmosis and babesiosis.

Treatment: Use the same insecticides as for lice. Treat the animal every two weeks if they have a problem. If the problem is bad, move them to a separate area after treatment.

Mange: (Sarcoptes) This is a microscopic anthropod that burrows under the skin and makes the animal scratch continually. They can greatly affect productivity of the animal.

Treatment: Use the same insecticides as for lice and treat all animals in the herd. Treat them every week for two or three treatments.

Maggots: These are fly larvae that can enter any type of wound, umbilical cord, or dirty, moist area. The flies lay their eggs on the wound and when they hatch, they enter the wound. Screw worms will continue to eat live flesh and enlarge the wound.

Treatment: When the animal has a wound spray it with insecticide such as Catron III three times per week until it heals. Watch goat kids to make sure maggots don't attack the umbilical cord. If you don't have insecticide spray, you may mix 1 quart of oil with 4 tablespoons of 57% liquid Malathion. This may be applied around wounds to treat or prevent maggots. Don't use too much, because this is poison. If the wound already has maggots, clean the wound out with water and a disinfectant.

Maggots can enter the gums, the umbilicus, vulva, or any wound or abscess.

 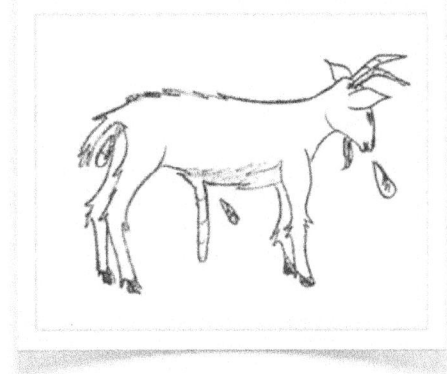

Mix: 1 quart of oil with 4 tablespoons of liquid Malathion or use insecticide spray.

Apply it to the wound

General Problems

Weak kids: Two to five days after birth, the kids can get too weak to nurse. This can happen if the kids do not get colostrum and bacteria can attack them. Without colostrum the kids can not fight off the bacteria and it is very hard to save them.

Treatment: Give them 1/2 cc penicillin once a day for 6 days or 1/2 cc LA 200 every three days for 6 days. Help them nurse or give them the mother's milk from a bottle. If the kid has diarrhea, give it electrolytes orally (1

teaspoon salt and 3 teaspoons of sugar per 1 coke bottle of water) each time it has diarrhea. Use a clean bottle.

Diarrhea: There are several things that can cause diarrhea: worms, bacteria, Coccidiosis, viruses, and certain plants and poisons. If the kid does not get colostrum the first day, it will likely get diarrhea.

Treatment: Remove any plants, feed, or poison that may have caused the problem. No matter what the cause, give oral electrolytes. (dosage above)

Wormer: If the kid is more than a month old, give worming medicine. See dosage on page 6.

Antibiotics: Give the kid sulfadiazine-trimethoprim (5 mg/pound twice a day). This is good for both bacteria and may help with Coccidiosis.

Abscesses: This is a pocket full of pus, usually just under the skin.

Caution: If you use the same needle on another animal that you used on a goat with abscesses, it may also get abscesses.

Treatment: Wait until the abscess is soft and then lance it large enough so the pus can drain out. Clean it out with water and disinfectant, then spray it with insecticide to prevent maggots from entering. Don't perform this procedure where the pus will drain on the ground where other goats graze.

The pus is full of bacteria and will cause problems in the other goats.

Remember, after you lance the abscess, the goat may still carry the bacteria in its throat and can still pass it on to other goats.

Abscesses

Each circle is a place where goats get abscesses most often.

When the abscess is soft:

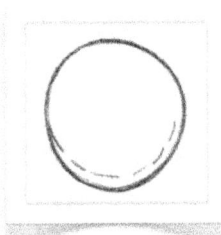

· Open the abscess towards the bottom and large enough for it to drain.

· Clean it well with water and disinfectant.

· Spray it with insecticide. (Screw-worm spray)

· Don't let the pus fall where there are other goats.

· It is best to burn the pus so it won't contaminate the other goats.

DO NOT BUY A GOAT WITH AN ABSCESS

Goats with abscesses can give it to the other goats. Even if you treat the abscess, the goat may still have more abscesses it in it's throat and contaminate the other goats.

Contagious Ecthyma: This is caused by a virus and may cause sores on the mouth or udders of the goats.

Treatment: There is no treatment for the virus, but you can use antibiotic ointment on the sores to help prevent bacterial infection. The sores on the mouth cause pain and the kids may not nurse. You may have to hand feed the kids if this happens. The sores on the udder may also hurt and the mother won't let the kids nurse. Once this goat has this once, it will never have it again. Don't purchase a goat and mix with your herd that has contagious ecthyma.

Caution: Be careful not to handle the sores without gloves this virus can cause the same sore in humans.

Sores on the udder of the doe.

Sores on the mouth of the kid.

Sores on a person's hand.

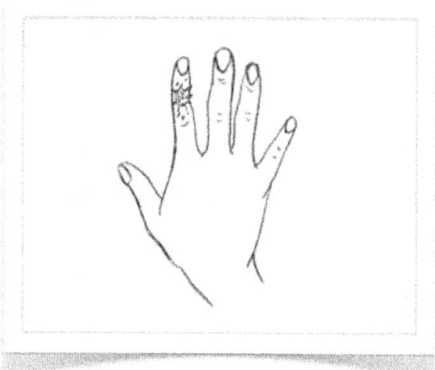

Abortions: A doe can abort if the feed is lacking in nutrition, if she is butted by another goat, if she eats some type of poison, or by some types of microorganisms.

Treatment: If you have only one goat that aborts, give her antibiotics for three days afterwards. Give her good feed and breed again when she is in good shape.

If several goats abort you need to find the cause. Check the feed to see if it is good. Also check to see if there is a goat that is fighting the others. After that give all your goats antibiotics to prevent further abortions. If a laboratory is available, check with it to see what samples may be sent for tests.

Dose of antibiotics:

Give 1 cc procaine penicillin per 30 pounds daily for three to six days.

or

Give 1 cc benzathine penicillin per 30 pounds every two days for 6 days.

or

Give 1 cc LA 200 per 20 pounds every 3 days for six days.

or

Give 5 mg Terramycin per pound daily for four to six days.

Separate the does that have not aborted from those that have and move them to a new area.

Metritis: This is an infection of the uterus. This can occur if the goat does not shed all of the placenta or if they shed the placenta within four hours after kidding or if they have aborted. Bacteria entering into the uterus cause infection. There may be pus and bad odor coming from the vulva.

Treatment: Give antibiotics such as Penicillin or Tetracycline for four to six days at the same dose as for abortions.

MASTITIS

Mastitis: This is an infection of the udder because of a wound or because bacteria entered the udder through the teat. When a goat has mastitis, the udder will be swollen and hot. The milk may be thick and clabbered or may look like water. The udder may also dry up. The doe will have a fever. If the udder is very sore, don't let the kid nurse.

Treatment: It is very important to milk out the udder several times per day until it heals. *Give the doe antibiotics such as Penicillin or LA 200 at the same dose as for abortions.*

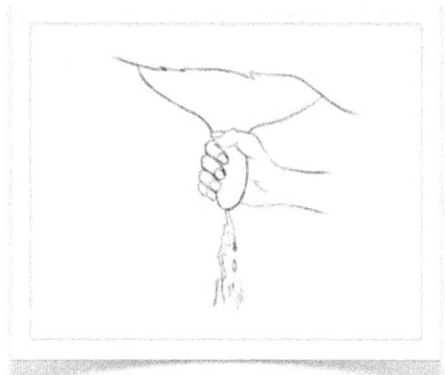 

Infection of the umbilicus and joints: When this happens, the umbilicus and joints will swell and get hot. The kid will be lame and have a fever. This happens because of bacteria entering the umbilical cord when the kid was just born. The bacteria then pass throughout the body and do much damage.

Treatment: If you don't move quickly to treat the umbilical infection, the bacteria will quickly move to the joints. Give 1 cc procaine penicillin per 30 pounds daily for 5 days or 1 cc benzathine penicillin per l0 pounds every 2 days for 6 days. Open the umbilicus so that it can drain and spray it for maggots. Prevention: Keep the kids where it is clean and dry. Dip the umbilical cord in disinfectant such as iodine or alcohol when they are born.

Joints swollen because of infection. Umbilicus swollen because of infection.

Arthritis: Problem in the joints because of infection or old age.

Treatment: There is not a good treatment for this. There is medicine for the pain such as aspirin.

Joints swollen because they are infected or the joints are large because of old age.

Infection of the foot (Foot Root):

When this happens the animal will be lame or refuse to move. The foot will be swollen between the hooves.

Treatment: Trim the hoof and look for a place where something may have pierced the foot. If there is something in the foot, remove it and put the animal in a place that is dry and clean and give penicillin or LA- 200. If you find an abscess, open it and clean out the pus. Then soak the foot in a disinfectant such as Clorox (3 parts water to 1 part Clorox) for 20 minutes daily. Give penicillin or LA-200.

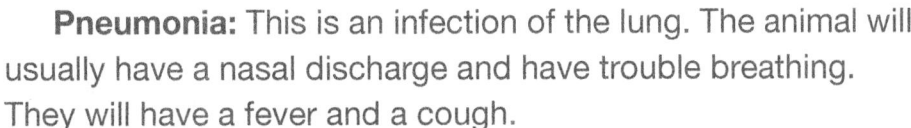

Trim the hoof and remove any object that may have stuck in the foot. Keep the goat in a dry, clean place.

Pneumonia: This is an infection of the lung. The animal will usually have a nasal discharge and have trouble breathing. They will have a fever and a cough.

Treatment: Give the animal penicillin or tetracycline until it is well. If you don't get improvement after two days, change the antibiotic. Use the same dose as for abortions (page 45).

Prevention: Keep the goats where they have fresh air. Don't keep them in a dirty, dusty pen. Have a shelter where they can stay out of the rain.

Tetanus: This is a deadly disease in goats that is caused by the tetanus organism entering a wound. The animal's body will be rigid and they can not move. The ears will be erect and they will jump when they hear noise.

Treatment: Give a triple dose of penicillin quickly (1 cc/10 pounds). Open and clean the wound with disinfectant. If you can find it, give 1500 units of tetanus antitoxin daily. Treatment is usually not successful.

Prevention: If you live in an area that has tetanus, you should vaccinate with tetanus toxoid. Procedure is given on page 6.

Rabies: Rabies can affect any mammal. Goats with rabies may first be weak in the rear legs and continually make a strange low bleating sound. It can not eat or drink. It may also act disoriented and sometimes be mean. If the goat bites a person or comes in contact with its saliva, they can also get rabies. The person should seek medical treatment quickly and the animal's brain sent to a laboratory for testing.

Treatment: There is no treatment for rabies. Don't eat the meat from an animal you suspect has rabies. The animal should be buried.

Prevention: All cats and dogs in the area should be vaccinated for rabies annually. Animals bitten by a mad dog or wild animal should be destroyed immediately. If you are in an endemic area for rabies, the goats can also be vaccinated with vaccine that is approved for domestic animals.

Anaplasmosis: This is a blood parasite of goats and cattle. The tick is the carrier of Anaplasmosis. It sucks blood from an animal that has the disease and then can transmit it to another. When a goat has Anaplasmosis, it becomes weak, depressed and anemic. It will also have a fever. The mucous membranes may be pale but most often yellow. `Animals that survive may remain carriers.

Treatment: Give the goat Tetracycline (9 mg/pound daily for five days). LA 200 is best since it only needs to be given every 3 days. You may also give B-complex to help build the blood.

Prevention: Treat the goats for ticks. If there are many ticks, you may need to move them to a new area after treatment (see page 40).

Babesiosis: This is also a blood parasite like Anaplasmosis and is also transmitted by ticks. When a goat has babesia it is depressed, weak, and thin. It will have a high fever and may have blood in the urine. The mucous membranes will be pale because of anemia. The animal may also have neurological signs and act crazy. The difference in signs of Anaplasmosis is that there will be blood in the urine. With Anaplasmosis the membranes will usually be yellow. Diagnosis can be made by microscopic identification.

Treatment: The best treatment for Babesiosis is Acaprine 5%. The dosage is .1 cc per 10 pounds subcutaneously. Caution: You must know the weight of the animal because Acaprine is extremely toxic and an overdose may kill the animal.

You may want to also give tetracycline. Babesia and Anaplasmosis can occur at the same time. B-complex will also be helpful in building the blood.

Prevention: Treating the animals for ticks in a Babesia endemic area is essential.

Animals that survive Babesia can carry the parasite in their blood and be a reservoir of the disease. Ticks feeding on a carrier animal may then pass it on to the other animals.

You can give a goat liquid medicine from a bottle like this.

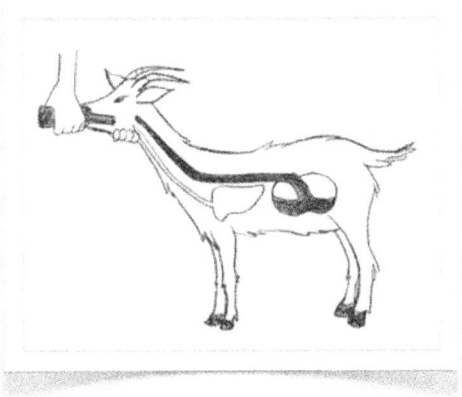

In the rear leg and in the neck.

There are two places you can give a goat shots in the muscle.

Kiling Goat For Meat

Hit the goat in the middle of the head just above the eyes so that he will be unconscious when you cut its throat.

Cut the goats throat so the blood will drain freely.

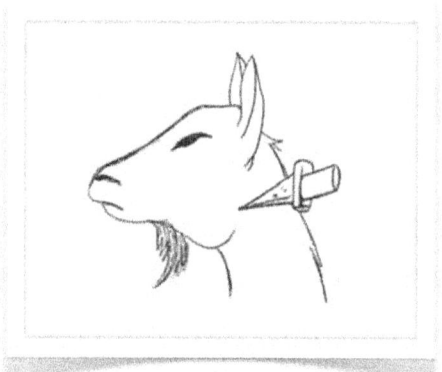 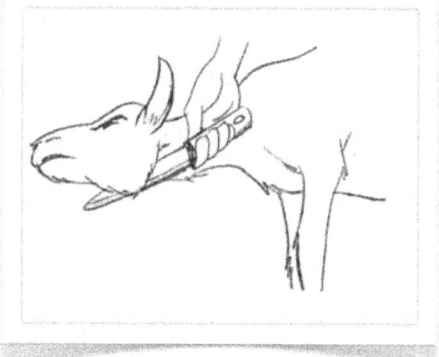

Hang the goat upside down so the blood will drain well.

If you do not bleed the goat quickly the meat will spoil faster.

Acknowledgment

RAISING GOATS

by Maureen E. Birmingham DVM

I thank God for the chance to work in the country of Haiti to prepare this book.

I also thank Christian Veterinarian Mission, CEDEPLA, IICA Save the Children and La Presse Evangelique for all the help they gave to help prepare this book.

I would also like to thank Mark and Peggy Rutledge for the big financial help they gave so the book could be printed.

Introduction

There is a big difference in people that just watch goats and people that raise goats. This book is for people that raise goats. It is them that want to make improvements so they can make more profit from raising goats.

It is my hope that this book can benefit those of you who raise goats and you can make progress.

I hope that after you study and apply what is in this book that you will help other people to do a better job of raising goats.

If you have any recommendations to make this book better, please send them to CVM, 19303 Fremont Ave. N., Seattle, WA 98133 206 546 7201

This book is dedicated to Dr. Leroy Dorminy founder of Christian Veterinary Mission Raising Goats

Let's Review

Read Section I and Section II

Fill in the correct word

4 days	dewormer	water
shelter	3-6 weeks	1 month old
two months	infection	deformation
tetanus vaccine	three	mud or rain
drained	forage	

1. Keep the goats in a place that is clean, dry and has _____.

2. Give the goats clean fresh _____ to drink.

3. Give the goats _____ that is green and young along with what other feed you can find.

4. Give the goats _____ every _____ months.

5. Don't drink the milk for _____ after giving the dewormer.

6. If there is a problem with tetanus in the area, give the goats _____ every year.

7. Trim the hooves often to prevent _____ and _____.

8. Goats do not like _____ or _____. They can get depressed.

9. Goats need a place that is well _____.

Answer the following questions

How do you prevent problem with worms?

10.

11.

12.

Read Sections III, IV and V of the Training Pac

Fill in the correct word

breeding	1 to 3	5 and 10
kids	5 months	antibiotics
one to two hours	placenta	21

13. Goats come into heat between _____ months.

14. It helps to feed the doe extra for three weeks prior to _____ to improve chances for more _____.

15. Goats come into heat every _____ days and stay in heat for _____ days.

16. The doe takes _____ to kid after breeding.

17. If everything goes well, the doe will kid in _____ hours after it starts labor.

18. If the doe had trouble kidding, give her _____.

19. Most does will shed the _____ within 4 hours time after kidding.

Answer the following questions

List at least 6 signs showing that delivery is close:

20.

21.

22.

23.

24.

25.

List the signs of labor and Normal Delivery:

26.

27.

28.

29.

30.

31.

What are the normal positions for delivery?

32.

33.

34.

What are the causes of Dystocia?

35.

36.

How do you help a goat with dystocia:

37.

38.

39.

40.

41.

42.

43.

44. What does the doe shed within 4 hours time after delivery?

_____

45.

_____

46.

_____

Score exercises 1 – 46

DO NOT look back at the Training Pac while completing the Review.

Review #1
(5 points each answer)

1. _____ Keep the goats in a place that is clean, dry and has shelter.

2. _____ Give the goats forage that is green and young along with what other feed you can find.

3. _____ Give the goats wormer every three months.

4. _____ Trim the hooves often to prevent infection and deformation

5. _____ Goats do not need a place that is well drained.

6. _____ Goats come into heat every month.

7. _____ It helps to feed the doe extra for three weeks prior to breeding to improve chances for more kids.

8. _____ The doe takes 5 months to kid after breeding.

9. _____ If the doe had trouble kidding, give her antibiotics.

10._____ The kid should appear after an hour when the water breaks.

Underline the correct answer

11. When you see signs of labor put the doe in a place by herself that is (clean, dirty) and has straw.
12. The kid should be born within (2 hours, 3 hours) time after the water breaks.
13. After delivery, the doe should shed the (placenta, umbilical cord) within 4 hours after delivery.
14. The doe needs more (juice, water) to help produce milk.
15. Move the kid (inside, outside) the uterus so it can be delivered.
16. Wash the hands and (arms, feet) very good with soap and water.
17. Let the kids nurse as soon as possible so they can get (colostrum, calcium).
18. The doe needs more feed to give plenty of (water, milk).
19. When you feed (forage, grass), tie it up to put it in a feeder so it can stay clean.
20. Give the doe antibiotics for (5 to 6, 7 to 10) days.

Score exercises 1 – 20

Read Sections VI and VII of the Training Pac

Underline the correct word

1. When a kid is first born, dip the (umbilical cord, placenta) in a disinfectant such as iodine and alcohol.

2. Help the kid nurse as soon as possible, it is very important for the kid to get (colostrum, calcium) as soon as possible.

3. (Debut, Debud) the goats if they will be in a pen with others so they don't damage others with their horns.

4. It is best to (castigate, castrate) the male goats before they are 3 weeks old.

5. If there is a problem with tetanus in the area, the kids should be (vaccinated, vindicated) against tetanus at one month of age and again at two months.

6. Heat up a metal pipe (1/2 inch , 1/4 inch) in diameter.

7. After it is very hot, place it on the small buds for (15, 20) seconds.

8. Be careful not to burn it too much and damage the (brain, hair).

Read Section VIII and Section IX of the Training Pac

Answer the following questions

What are the two ways to castrate a goat?

9.

10.

Choose a Goat that is:

11.

12.

DO NOT choose a goat that has: (Give 6 answers)

13.

14.

15.

16.

17.

18.

Choose a doe that has:

19.

20.

21.

22.

23.

24.

A good buck has: (Give 6 answers)

25.

26.

27.

28.

29.

30.

Read Section XII of the Training Pac

Answer the following questions

When keeping a goat in a pen:

31.

32.

33.

34.

How to get the benefits of a pen:

35.v

36.

37.

38.

Fill in the blanks

39.Put the salt in a _____.

40.The salt will _____ through the gourd.

41.When you keep goats in a pen you must cut _____ and _____ it to them.

42. It is better to put the hay in a _____ or tie it up so the goat will have a clean feed that isn't _____ with _____ eggs.

43. Goats will not eat _____ feed.

44. It is best for the forage to be _____ and _____ to make the best hay.

Fill in the correct word.

Spread	tastier	dry season
clear	shelter	young
green	dew	sun
spoil	storage	salt

45. When you make hay, you will conserve it for the _____ _____.

46. Cut the forage for _____ the time that is _____. Cut it after the _____ has dried.

47. _____ the forage and turn it daily so the _____ can dry it.

48. Put the hay inside or under a _____ so they won't get wet, because if it gets wet it will _____.

49. To make the hay _____, put a small amount of _____ between each layer.

Answer the following questions.

How can you lose hay?

50.

51.

52.

53.

Score exercises 1 – 53

STOP!

DO NOT look back at the Training Pac while completing the Review.

Review #2
(5 points each answer)

Underline the correct word

1. When a kid is first born, dip the (umbilical cord, placenta) in a disinfectant such as iodine and alcohol.

2. (Debut, Debud) the goats if they will be in a pen with others so they don't damage others with their horns.

3. It is best to (castigate, castrate) the male goats before they are 3 weeks old.

4. Heat up a metal pipe (1/2 inch , 1/4 inch) in diameter.

5. Be careful not to burn it too much and damage the (brain, hair).

6. Choose a goat that is (not too fat, too fat).

7. Do not choose a goat that is too(old, young).

8. Choose a doe that has a good (other, udder).

9. Choose a buck that has (strong and straight, long and narrow) legs.

10. Choose a buck that has good (teeth, dentures).

True or False

11._____ When you keep a goat in a pen you can control the breathing.

12._____ To get the benefits of a pen you must give the goats fresh clean water everyday

13._____ You must plant forage for the goat.

14._____ You must give worming medicine at least every other month.

15._____ When you keep goats in a pen you must cut forage and carry it to them.

16._____ Goats will not eat dirty feed.

17._____ It is best for the forage to be brown and old to make the best of hay.

18._____ Spread the forage and turn it daily so the sun can dry it.

19._____ Put the hay inside or under a cave so the hay won't get moldy.

20._____ The hay will lose its leaves if it is too dry.

Score exercises 1 - 20

Read Section XIII of the Training Pac

Fill in with the correct word

loose	yoke	vegetation	destroy	trees

1. Use living _____ for post.

2. Use cactus or prickly _____.

3. You must use a _____ to help them from getting out.

4. Goats running _____ will kill small trees and _____ gardens.

Read Section XIV and Section XV of the Training Pac

Answer the following questions:

What are the signs of sickness or disease in a goat? (Give at least 10 answers)

5.

6.

7.

8.

9.

10.

11.

12.

13.

14.

What different types of Internal Parasites can goats get?

15.

16.

17.

18.

What Different types of External Parasites can goats get?

19.

20.

21.

22.

What are the General problems goats can have? (Give at least 6 answers)

23.

_____

24.

_____

25.

_____

26.

_____

27.

_____

28.

_____

Read Section XVI of the Training Pac

Fill in the blanks

upside down	blood
drain	middle
meat	spoil
unconscious	

29. Hit the goat in the _____ of the head just above the eyes so that he will be _____ when you cut his throat.

30. Cut the goat's throat so the _____ can _____ freely.

31. Hang the goat _____ so the blood can drain good.

32. If you do not bleed the goat quickly the _____ can _____ faster.

Score exercises 1 - 32

STOP!

DO NOT look back at the Training Pac while completing the Review.

Review #3

(10 points each answer)

Matching

a.trees	f.head
b.destroy	g.lice and mange
c.vegetation	h.internal parasites
d.yoke	i.meat can spoil faster
e.anemic	j.a general problem

1. You must use a _____ to help keep the goats from getting out

2. Use cactus or prickly _____

3. Use living _____ for post

4. Goats running loose will kill small trees and _____ gardens

5. A goat that is sick can have a large

6. A goat that is sick can be

7. Lung worms are

8. External parasites

9. Abscesses

10. If you do not bleed the goat quickly

Pre-Test

(4 points each answer)

True or False

1. _____ Keep the goats in a place that is clean, dry and has shelter.

2. _____ Give the goats forage that is green and young along with what other feed you can find

3. _____ Trim the hooves often to prevent infection and deformation

4. _____ Goats do not need a place that is well drained.

5. _____ Goats come into heat every month.

6. _____ The doe takes 5 months to kid after breeding.

7. _____ If the doe had trouble kidding, give her antibiotics.

8. _____ When you see signs of labor put the doe in a place together with other goats that is clean and has straw.

9. _____ The kid should be born within 2 hours time after the water breaks.

10. _____ The buck needs more water to help produce milk.

Fill in the correct word

castrate	debud
udder	strong and straight
umbilical cord	not too fat
1/2 inch	teeth

11. When a kid is first born, dip the _____ in a disinfectant such as iodine and alcohol.

12. _____ the goats if they will be in a pen with others so they don't damage others with their horns.

13. It is best to _____ the male goats before they are 3 weeks old.

14. Heat up a metal pipe _____ in diameter.

15. Choose a goat that is _____.

16. Choose a doe that has a good _____.

17. Choose a buck that has good _____.

18. Choose a buck that has _____ legs.

Matching

a. forage	e. fresh clean water
b. anemic	f. vegetation
c. trees	g. lice and mange
d. breeding	

19. When you keep a goat in a pen you can control the _____.

20. To get the benefits of a pen you must give the goats _____ everyday.

21. You must plant _____ for the goat.

22. Use cactus or prickly _____

23. A goat that is sick can be

24. External parasites

25. Use living _____ for post

Score exercises 1 – 25

STOP!

You must now prepare yourself for the Final Test. In preparation, you may want to follow one or `more of these suggestions:

1. Review the Contents.

2. Review every incorrect exercise in the Pre-Test.

3. Reread each section of the Training Pac.

4. Relearn each section you still do not completely understand.

Let's Review Score Key

Section I and Section II

1. Shelter

2. Water

3. Forage

4. Dewormer...three

5. 4 days

6. Tetanus Vaccine

7. Infection...deformation

8. Mud...rain

9. Drained

10. Give dewormer every three months.

11. When you feed forage, tie it up or put it in a feeder so it can stay clean.

12. Stake the goats in clean areas that have not been contaminated by other goats change the place everyday.

Sections III, IV and V

13. 5 and 10

14. breeding...kids

15. 21 days...1 to 3 days

16. 5 months

17. One to two

18. Antibiotics

19. Placenta

20. The doe stays by herself.

21. She is nervous.

22.She paws the ground

23.There is a thick mucous discharge from the vulva .

24. The udder is full. Colostrum may wax the end of the teat.

25.She lays down often.

Or She looks at her side.

Or She is uneasy.

Or She bleats a lot

Or The stomach drops

Or There is a depression.

26.When you see signs of labor put the doe in a place by herself that is clean and has straw.

27.Her water breaks

28.The kid should appear within an hour after the water breaks.

29.The kid should be born within 2 hours time after her water breaks

30.Dip the navel cord in disinfectant.

31.Let the kids nurse as soon as possible so they can get colostrum.

32.Position of the kid: The head is in between the front legs facing forward. The front legs come out first.

33.Rear legs come out first with the back up.

34.Both kids are in normal position in the uterus.

35.Rear of the kid coming first.

36.The head turned back.

37.Cut the fingernail very short.

38.Wash the hands and arms very good with soap and water.

39.Wash the rear of the doe with soap and water.

40.Push your hand inside very gently.

41.Move the kid inside the uterus so it can be delivered.

42. Pull the kid by the front legs. The front legs should come out before the head.

43. Give the doe antibiotics for 5 to 6 days.

44. The placenta

45. The doe need more water to produce milk.

46. The doe need more feed to give plenty of milk.

Review #1

1. T

2. T

3. F

4. T

5. F

6. F

7. T

8. T

9. T

10. F

11. clean

12. 2 hours

13. placenta

14. water

15. inside

16. arms

17. colostrum

18. milk

19. forage

20. 5 to 6

Section VI and Section VII

1. placenta

2. colostrum

3. Debud

4. castrate

5. vaccinate

6. 1/2 inch

7. 15

8. brain

Section VIII and Section IX

9. You can crush the cord

10. You can cut the scrotum or remove the testicles.

11. Not too fat

12. Not too thin

13. Weak back

14. Too old

15. Check teeth narrow nose and mouth

16. Straight shoulders

17. Flat stomach

18. Weak legs and fetlocks

Or Sagging udder
Or Rump that is too sloped and weak

19. Good teeth

20. Strong, straight back

21. Stomach that has a large capacity

22. Strong legs

23. Good udder

24. Legs that are straight

25. Clear eyes

26. Large head with a long mandible

27. Large mouth and nose

28. Strong and straight legs

29. Heavy muscled

30. Large and strong body

Or Strong back with long rump
Or Large testicles that are the same size Or Good teeth

Section XII

31. You can control the breeding.

32. You can watch the goats better.

33. You must find feed for the goats.

34. It is easier to give clean water and good hay.

35. You must give them fresh clean water everyday.

36. You must plant forage for the goats.

37. You must conserve forage for the dry season. This is called hay.

38. You must give worming medicine at least every three months.

39. Gourd

40. Foak

41. Forage...carry

42. Feeder...contaminated...worm

43. Dirty

44. Green...young

45. Dry season

46. Storage...clear...dew

47. Spread...sun

48. Shelter...spoil

49. Tastier...salt

50. If it is stacked too wet, it will get hot and spoil.

Final Test

FINAL TEST
(4 points each answer)

True or False

1. _____ Keep the goats in a place that is clean, dry and has shelter.

2. _____ Give the goats wormer every three months.

3. _____ Trim the hooves often to prevent infection and deformation.

4. _____ Goats do not need a place that is well drained.

5. _____ The doe takes 7 months to kid after breeding.

Fill in the correct word

   (a) not too fat   (b) good udder   (c) iodine and alcohol

   (d) teeth   (e) 3 weeks old

6. The doe needs more _____ to help produce milk.

7. After delivery, the doe should shed the _____ within 4 hours after delivery.

8. Wash the hands and arms very good with _____.

9. Let the kids _____ as soon as possible so they can get colostrum.

10. The doe needs more _____ to give plenty of milk.

Matching

(a) not too fat   (b) good udder   (c) iodine and alcohol

(d) teeth   (e) 3 weeks old

11. When a kid is first born, dip the umbilical cord in a disinfectant such as
_____.

12. It is best to castrate the male goats before they are _____.

13. Choose a goat that is _____.

14. Choose a doe that has a _____.

15. Choose a buck that has good _____.

Underline the correct word

16. When you keep a goat in a pen you can control the (breathing, breeding).

17. You must plant (forage, foliage) for the goat.

18. Goats will not eat (dirty, clean) feed.

19. Spread the forage and turn it (daily, quickly) so the sun can dry it.

20. To get the benefits of a pen you must give the goats fresh clean water
(everywhere, everyday).

Choose the correct answer

21. You must use a yoke to help keep the goats
(a) from getting out (b) from getting scared

22. Goats running loose will kill small trees and
(a) spoil the vines (b) destroy gardens

23. A goat that is sick can have
(a) a fever (b) a large head

24. A goat that is sick can be
(a) anemic (b) diabetic

25. If you do not bleed the goat quickly
(a) meat can spoil faster (b) meat can cool faster

Final Test Score Key

Final Test

1. Tuberculosis

2. Lice...ticks...fleas

3. Gut

4. Dengue

5. Tuberculosis

6. Tetanus

7. Rabies

8. Meningitis

9. Mosquitoes

10. Brucellosis

11. Food...water

12. Typhoid

13. Leprosy

14. Splinted

15. Hansen's Disease

CHAPTER 4

Improved Backyard Chicken Production

Objectives To know that raising backyard chickens are economical and are good for family nutrition as well as an excellent source of improving family income.

To successfully make a choice in selecting local or hatchery chickens and properly care for your backyard chickens.

To know whether a hen is broody as well as know how to have successful natural incubation.

Words to Know

Antibiotic: a medicine given to treat against disease.

Eliminate: to get rid of.

Hatchery: special chickens sold for the purpose of laying eggs or producing meat.

Incubation: the period of time it takes for chickens to hatch.

Infectious: the effect of disease producing germs causing an infection.

Management: to take charge or care for something.

Parasite: an animal that lives in or on a different kind of animal and gets its food from the animal, often injuring the animal.

Perch: a pole, bar or branch (usually lifted off the ground) that a bird may rest upon.

Predators: an animal or person that will take or harm the chickens or eggs.

Production: a product, such as eggs, chickens multiplying.

Susceptibility: sensitive, easily affected by.

Vaccination: a medicine given to an animal that will protect them from getting a disease.

Improved Backyard Chicken Production

Everyone can raise backyard chickens. Starting with one broody hen and twelve eggs, a person can increase to 50 adult birds in less than two years.

With backyard chickens there will always be eggs which are important for improving family nutrition. Because eggs are rich in protein, it is good to serve an egg daily to children so that they will grow strong and healthy.

Chicken and eggs are highly regarded foods and therefore a good market always exists, thus providing a way to improve family income.

Backyard Chicken Management

A farm family can usually produce enough corn to feed up to 50 adult chickens.

To obtain good chicken and egg production, it is necessary to learn the following simple management practices.

Local Breeds

Local chickens and their eggs are better tasting and have a higher market value when compared with hatchery chickens.

Local hens are excellent for hatching eggs and raising chicks. With local hens, one does not need to buy chicks nor expensive feeds from the hatchery. Local chickens can be raised in the backyard without much care, as they feed themselves by scratching the ground for insects and weeds.

To keep these valuable characteristics, local birds should not be crossed with hatchery birds.

Comparison of local and hatchery breeds

Point of Comparison	Local Breeds	Hatchery Breeds
Cost of chicks	Inexpensive	Expensive
Cost to feed	Inexpensive	Expensive
Death rate	Low	High
Type of growth	Hardy	Delicate
Egg laying	Average	Very good
Natural Incubation	Excellent	Poor
Raise Chicks	Excellent	Poor

Broody Hens

A broody hen lifts it feathers, makes a warning sound and seldom leaves it nest. These signs should be observed for two or three days before setting the broody hen with eggs. The best broody hens completely cover 10 – 12 eggs in the nest.

To break the broodiness of poor mothers, it is necessary to build a wire bottomed pen adjoining the chicken coop where the broody hen can be enclosed for several days.

Natural Incubation

To have successful natural incubation, one should select well-formed, medium to large sized eggs. Small eggs produce weak chicks. It is better to use eggs less than ten days old. For fertile eggs, there should be one rooster per 15 hens.

One hen can incubate 10 to 12 eggs. Before sitting the hen, mark the date on each egg with a pencil to avoid confusion with freshly laid eggs.

After setting the broody hen, keep the hen enclosed for the first day so that she can become used to the nest. Beginning on the second day, turn the setting hen loose each morning so that she can eat and drink. In an hour, remove any unmarked eggs from the clutch and re-enclose the hen on the nest.

The chickens should then hatch in 21 days. Before turning them loose, it is good to keep the mother hen and her baby chicks in the nest with sufficient ground corn and fresh water for two days. After this, the baby chicks are sufficiently strong enough to go outside with the mother hen.

Feeding Hens and Chicks

Corn is necessary feed for good egg production. Each morning one should give one pound of shelled corn for each ten adult hens. Also, it is good to feed tubers, bananas or spoiled fruit.

Give ground corn to baby chicks which can not eat whole grains. To give ground corn to baby chicks, it is good to build a slatted pen or creep feeder with spaces the width of two fingers. This creep feeder should be under the chicken coop roof to protect the feed and chicks. Sufficient ground corn for the whole day should be placed in the center of the slatted pen.

General Care

The chicken coop should remain open all day so that the hens can enter and lay eggs. Eggs should be collected frequently and stored in a cool place to prevent from spoiling especially in hot weather. Hatching eggs require special care 7 to 10 days, 13°C (55°F).

The chicken coop should be closed each night to keep out dangerous animals. Before closing the door, check to see if all the mothers are with chicks, especially those that went outside for the first time, have returned.

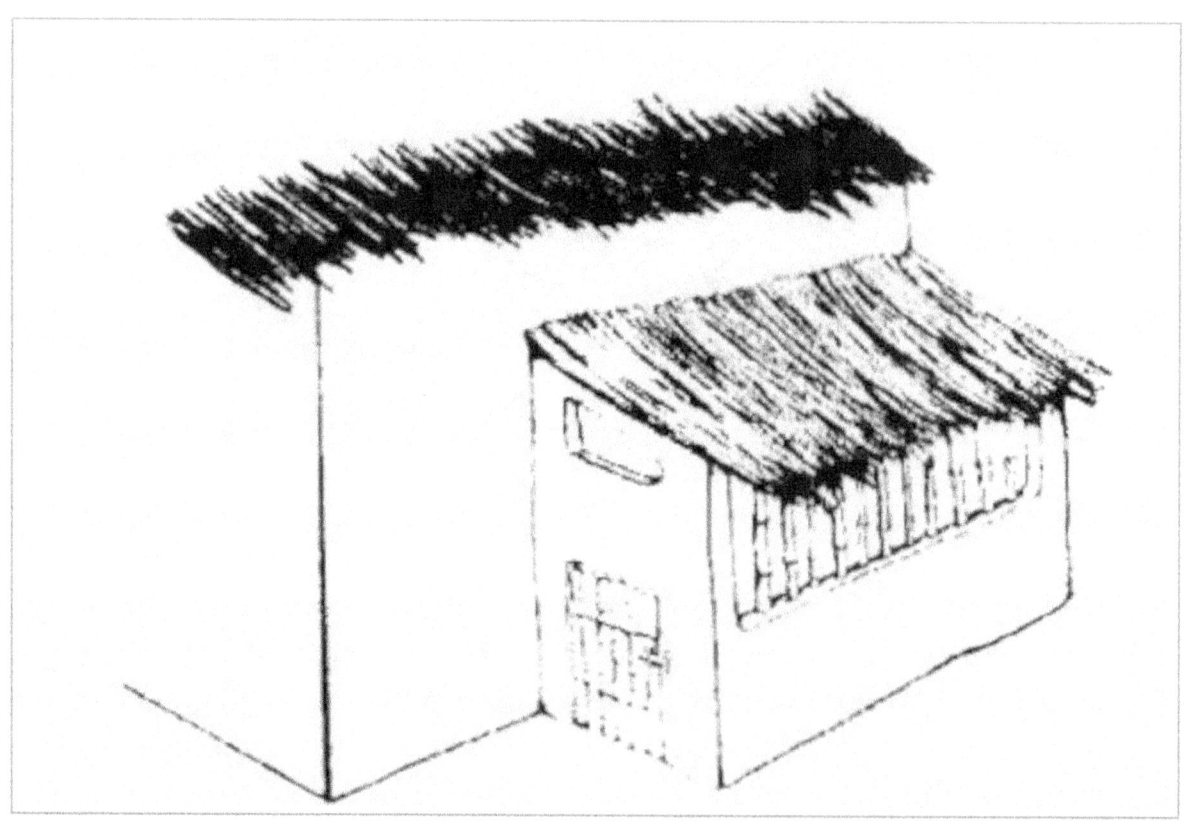

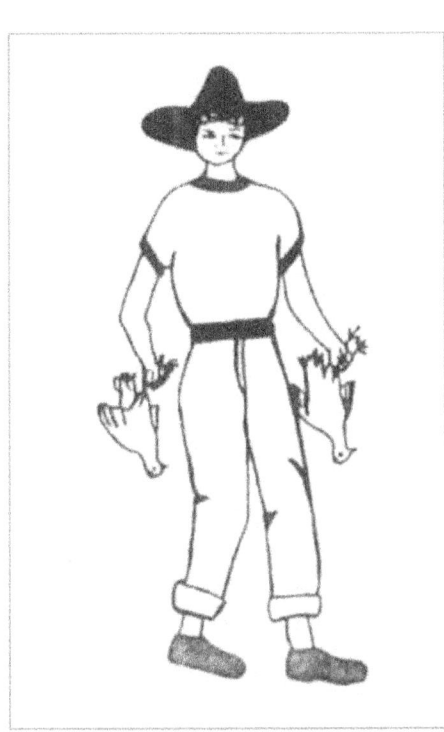

Sale of Birds

Excess young roosters and older birds that are no longer desirable for breeding should be eliminated each month. Older birds can be identified by their fallen crests, destroyed feathers and rough feet.

It is only necessary to maintain one rooster per 15 hens. Hens suffer when many roosters mount them.

Chicken Coop Construction

Chicken Coop Construction

The chicken coop should be strongly built so that night predators can not enter. One can build a chicken coop with boards from the trunk of palm trees, wooden poles and thatch material. A chicken coop of five meters is sufficient for 50 adult chickens.

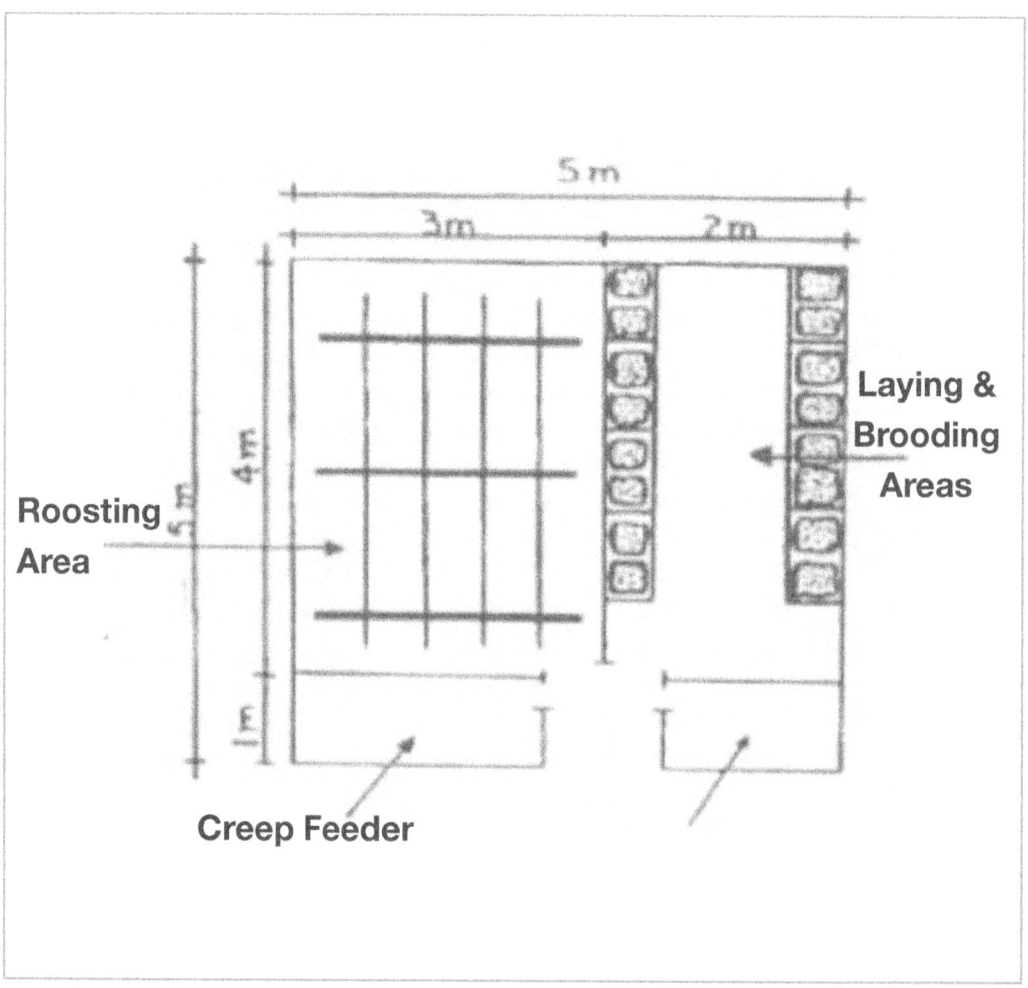

The chicken coop should have a large room with perches. It should also have a small room with laying boxes and nests for setting eggs. It is a good practice to fill the boxes and nests with chopped cornhusks. Change the litter often, keeping the coop clean and dry. Nests for the natural incubation can be built-in under the laying boxes. Mothers with small chicks sleep in the small room during the first few weeks.

Materials Needed	
4 – 2" x 2" x 8' (top/bottom rails)	A
2 – 2" x 2" x 6' (end rails)	B
7 – 2" x 2" x 6' (side rails)	C
8 – 2" x 2" x 4' (roost bars / braces)	D
Chicken Wire (1" mesh) 6 x 20' roll	
Exterior plywood 4 x 8' x 1/2" Nest boxes, closed end, corner braces	
Corrugated materials: 2 – 6' x 4	5 sheets needed to cover roof
Water for up to 15 birds	
Bird feeders	

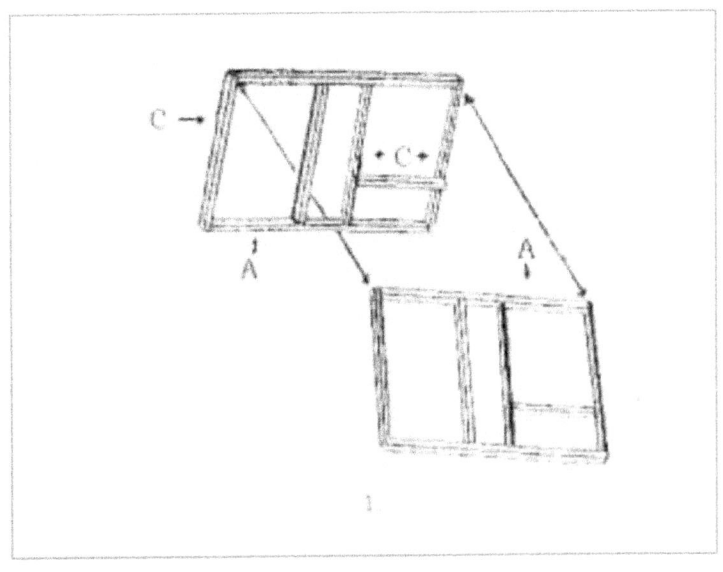

1. Build and connect side frames

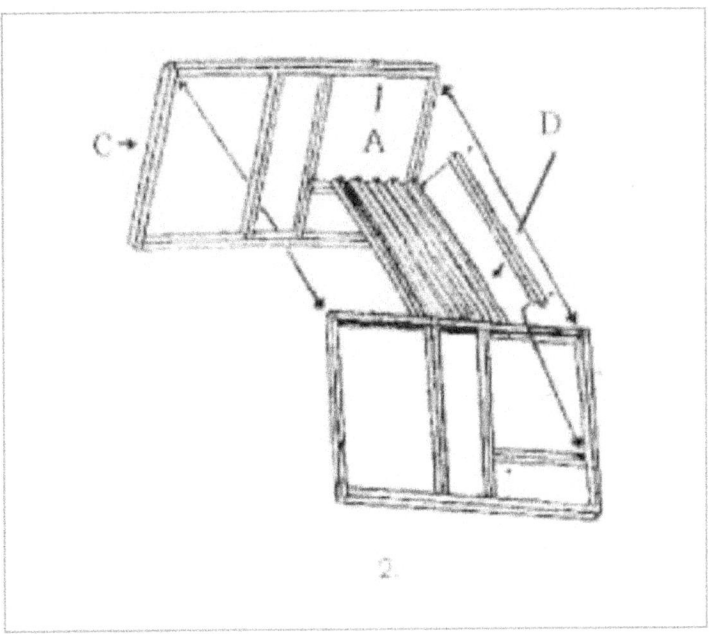

2. Build roost

3. Attach corner braces

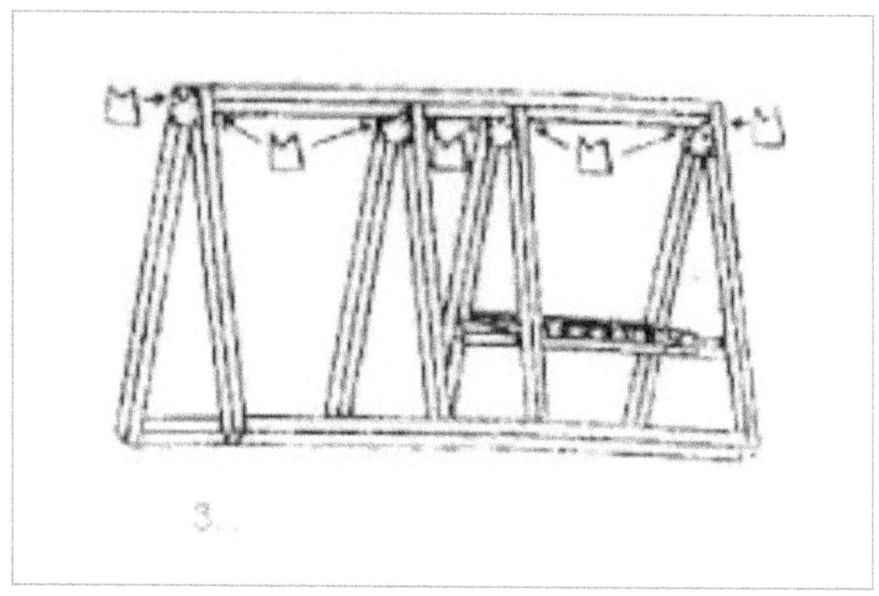

4. Finished frame

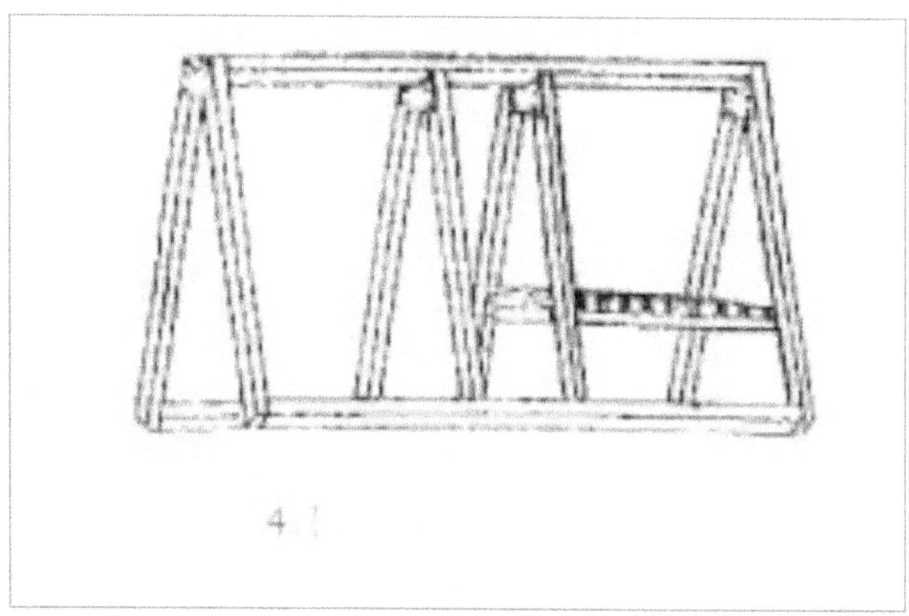

5. Attach closed ends, end rails and door

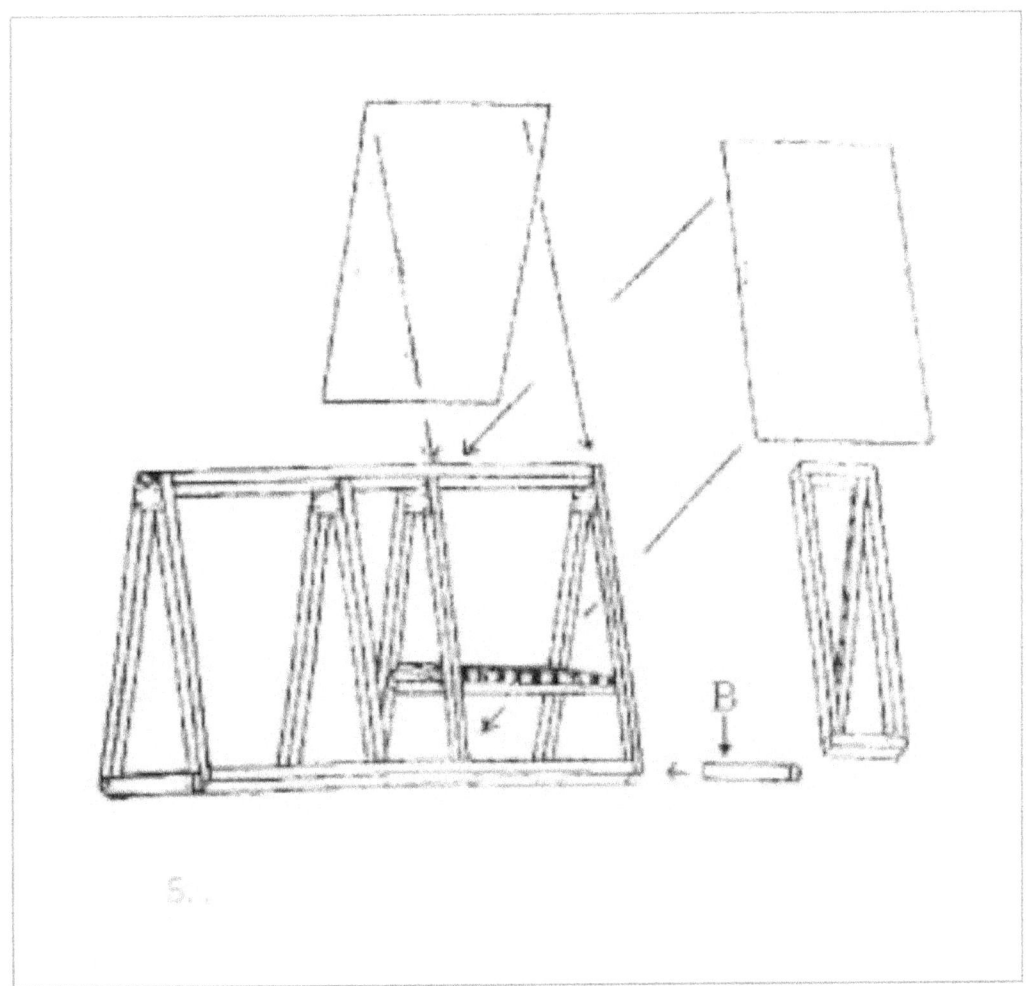

6. Add mesh to end and nest boxes

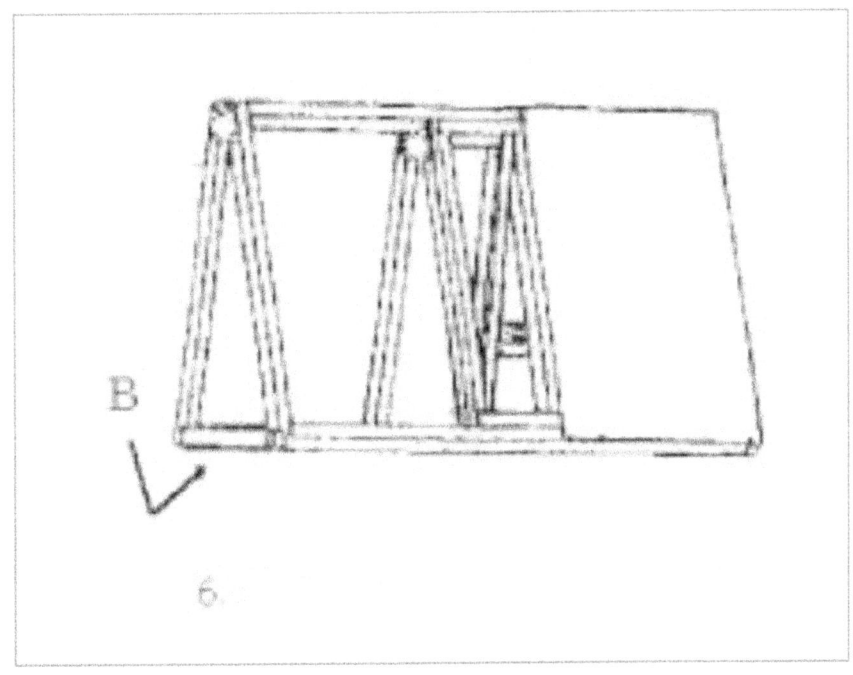

Basic Disease Control

Basic Disease Control

It is necessary to learn these simple practices of disease control to avoid the loss of chickens and time invested in them.

Vaccination against Newcastle Disease

By vaccinating against Newcastle one avoids the great loss to death this disease causes. Signs include gasping, coughing, and drooping wings. Almost all birds die after two or three days.

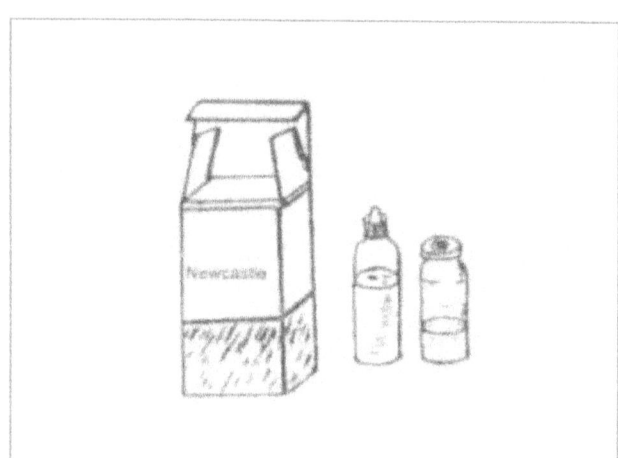

The vaccination is for the whole flock and needs to be repeated every three months. The vaccination is done by placing one drop in the eye being careful not to touch the eye with the applicator. The vaccine can be bought under the name NEWCASTLE in a veterinary drug store. As the vaccine comes in powdered form it can be kept without refrigeration for one week, but once mixed with the liquid it should be used the same day. When vaccinating, be careful not to place your hands near the face as the vaccine can infect human eyes.

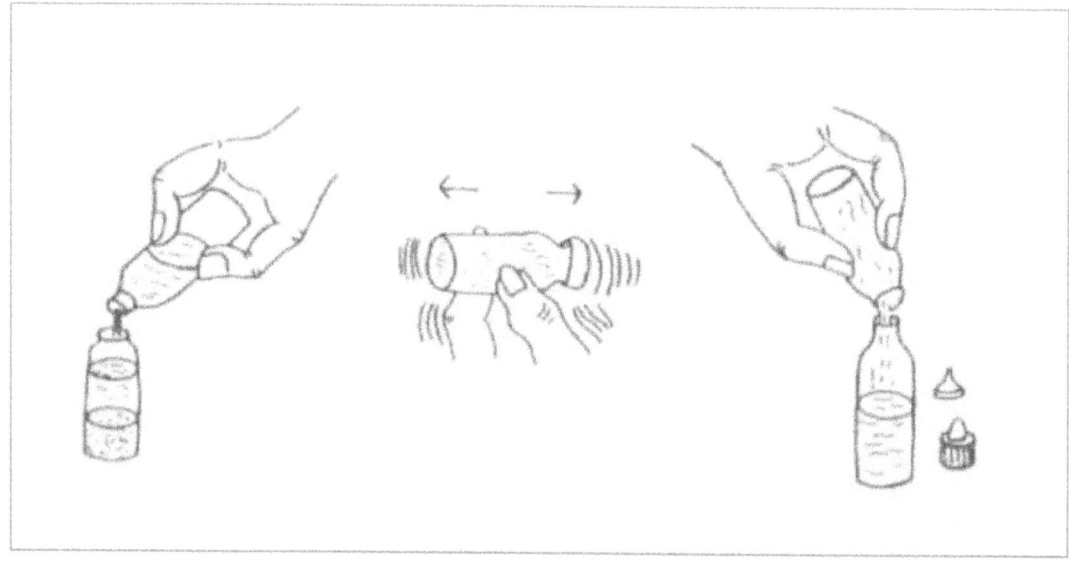

When finished with the vaccination, burn any remaining vaccine with the container and wash hands thoroughly.

Internal Parasite Control

Chickens will be stronger and healthier when parasites are controlled. By not controlling internal parasites, chickens will not grow as well, they will have lower egg production and increased susceptibility to other diseases. Small chickens are the most seriously affected by internal parasites and some may even die.

It is best to deworm the entire flock with pills according, to these sizes:

Chicks 1/2 pill

Adults 1 pill

Deworming pills can be purchased in veterinary drug stores under the name TRIPLE WORMER or WORMAL. The treatment should be repeated every three months. To save time, deworming can be done with the vaccination against Newcastle disease.

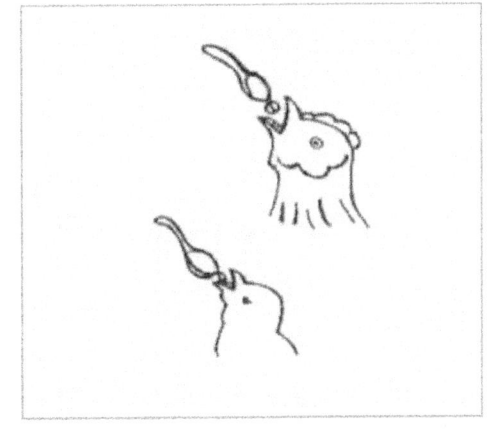

Lice Control

Lice should be controlled as they cause weight loss and lower egg production. Lice are also a very common problem on setting hens. Lice can be controlled by mixing one part MALATHION powder with four parts of ashes.

The mixture is easy to shake on using a jar with a top that has holes. The wings and legs of all birds should be dusted every three months, at the same time as the other control practices. In addition, hens and their nests should be dusted at the beginning of natural incubation.

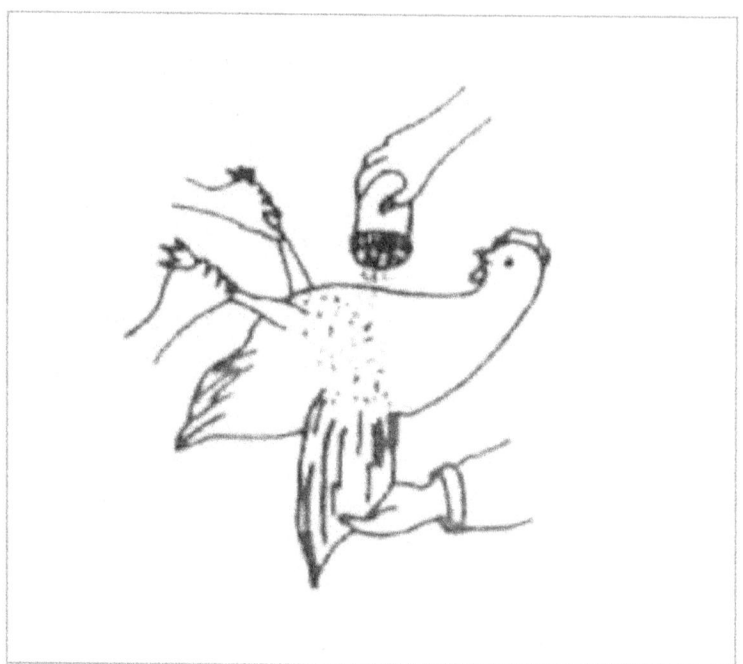

Control of Chronic Respiratory Disease

Chronic Respiratory Disease is a common long lasting infection characterized by nasal discharge. This nasal discharge causes difficulty in breathing, producing a rattling or bubbling sound. Sick birds lose their appetite and do not lay many eggs. Small chicks are the most seriously affected and some may even die.

Chronic Respiratory Disease can be controlled with the antibiotic Terramycin. For administration, a four gram package can be mixed in two cups of water.

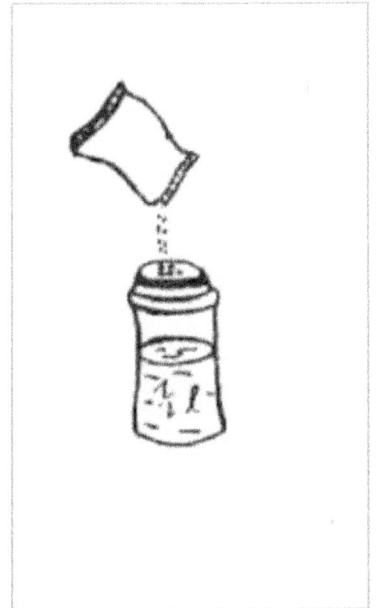
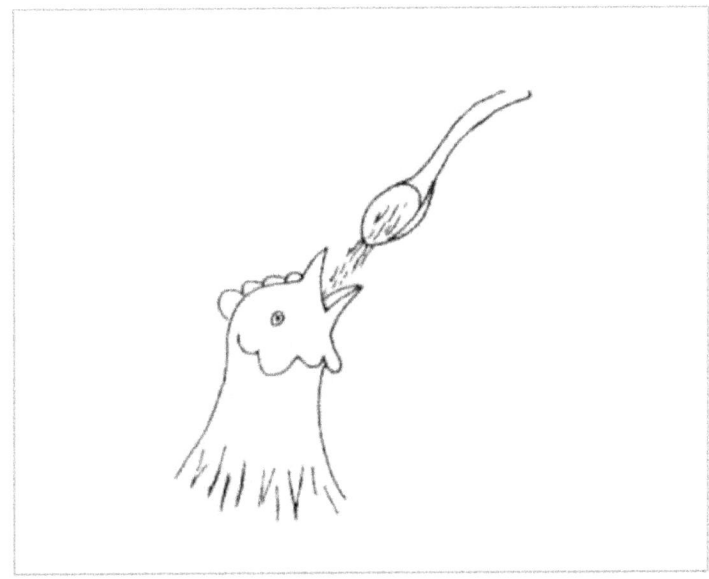

This quantity is sufficient to treat 100 birds. It is necessary to treat all sick and healthy birds, giving each a teaspoonful in the mouth. This treatment should be repeated every three months at the time as the other control practices.

Important Disease Control Practices to be Done Every Three Months

Newcastle

Internal parasites

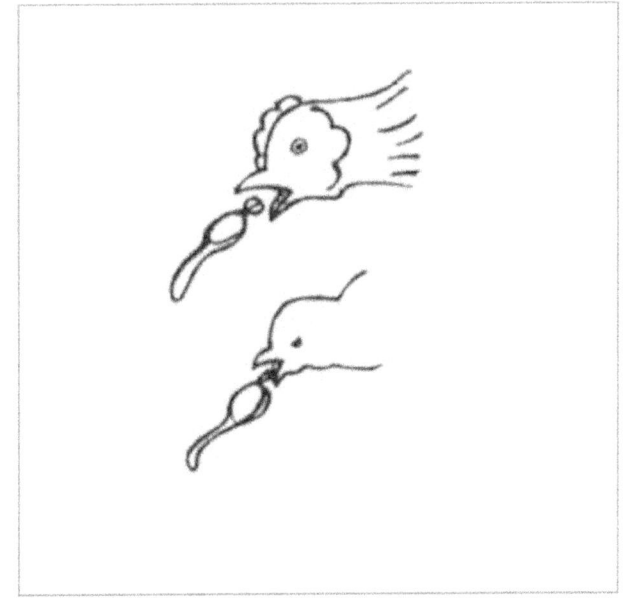

Pill in the mouth

Chronic Respiratory Disease	**Lice**

Antibiotic in the mouth	**Dust under the wings**

Suggested Recommended Schedule of Vaccination

BAI Philippines

Vaccination	Age of Bird
Avian Pest Vaccine (Intranasal method)	1 day – 1 week old
Pigeon Pox Vaccine	one month old
Roup Vaccine	two months old
Avian Pest Vaccine (prick method)	three months old
Fowl Pox Vaccine	four months old
Avian Pest Vaccine	repeat after 1 year of laying

Acknowledgments

Used by Permission

Authors:
Walter Baquero A.
Wimper Baquero A

.

Collaborators
Jose Martinez O.
Henry Oniate R.

Illustrators
Sheri L. Borman
Allyson J. Kramer

Advisors John P. Bishop, Kay P. Bishop

Published by the Summer Institute of Linguistics under contract
with the National Institute of Agricultural Research (INIAP) Limoncocha 1982
(Reproduction is authorized)

Linguistics Press Ecuador
"Be A "Good Farmer"
Agricultural Education Series
Second Book of Agricultural Education or Amazonian Ecuador

Let's Review

Improved Backyard Chicken Production

Fill in the blanks

Matching the word on the left with the definition on the right

1. ___ Antibiotic

2. ___ Hatchery

3. ___ Incubation

4. ___ Infectious

5. ___ Parasite

6. ___ Predators

7. ___ Susceptibility

8. ___ Vaccination

9. ___ Management

10. ___ Perch

a. The period of time it takes for chickens to hatch

b. An animal that lives in or on a different kind of animal and gets its food from the animal, often injuring the animal.

c. Special chickens sold for the purpose of laying eggs or producing meat.

d. An animal or person that will take or harm the chicken or eggs.

e. A medicine given to an animal that will protect them from getting a disease.

f. A medicine given to treat against disease.

g. The effect of disease producing causing an infection.

h. To take charge of caring for something.

i. A pole, bar or branch that a bird may rest upon.

j. Sensitive, easily affected by.

11. _____ can raise backyard chickens.

12. Starting with one _____ _____ and _____ eggs, you can increase to 50 adult birds in less than two years.

13. Because eggs are rich in _____, it is good to serve an egg daily to children so that they will grow _____ and _____.

14. _____ and _____ are highly regarded foods and provides a way to improve _____ _____.

15. A farm family can usually produce _____ to feed _____ adult chickens.

16. Local chickens and their eggs are better _____ and have a _____ market value when compared with _____ chickens.

17. Local hens are excellent for _____ eggs and _____ chicks.

18. Local chickens feed themselves by _____ the ground for _____ and _____.

19. _____ _____ should not be crossed with hatchery birds.

Answer the following questions

20-22. What are the signs of a broody hen:

20. _____

21. _____

22. _____

23. How many days should signs be observed before setting the hen with eggs?

24. How many eggs can a best broody hen cover?

25. What should one do to have a successful natural incubation?

26. How many roosters and hens do you need for fertile eggs?

_____

27. How many eggs can one hen incubate?

_____

28. What should you do before sitting a hen?

_____

29. How many days do chickens hatch?

_____

30. When you keep a mother hen and her baby chicks in the nest what do they need for two days?

_____

Underline the correct word

31. Give (ground, grilled) corn to baby chicks which cannot eat whole grains.

32. The chicken coop should remain (open, close) all day so that the hens can enter and lay eggs.

33. Hatching eggs require special care (7 to 10 days, 3 to 5 days).

34. The chicken coop should be closed each night to (keep, keep out) dangerous animals.

35. Excess young roosters and other older birds that are no longer desirable for breeding should be (qualified, eliminated).

36. It is necessary to maintain one rooster per (10, 15) hens.

37. The chicken coop should be strongly built so that the night (predators, owls) can not enter.

38. A chicken coop of 5 meters is sufficient for (50, 100) adult chickens.

39. It is a good practice to fill the boxes and nests with chopped (coconut husk, cornhusks).

Review #1

Fill in with the correct word

Antibiotic	Hatchery
Incubation	Predators
Vaccination	

1. _____ is a medicine given to an animal that will protect them from getting a disease.

2. _____ is the period of time it takes for chickens to hatch

3. _____ special chickens sold for the purpose of laying eggs and producing meat.

4. _____ is a medicine given to treat against disease

5. _____ an animal or person that will take or harm the chickens or eggs.

Respond with a T for those answers that are TRUE and with an F for FALSE

6. ___ Only farmers can raise backyard chickens.

7. ___ Starting with one broody hen and twelve eggs you can increase to 50 adult birds in less than two years.

8. ___ Because eggs are rich in protein, it is good to serve an egg every other day to children.

9. ___ A farm family cannot produce enough corn to feed up to 50 adult chickens.

10. ___ Local hens are excellent for hatching eggs and raising chicks.

11-13. What are the signs of a broody hen?

11. _____

12. _____

13. _____

14. How many eggs does a best broody hen can completely cover?

_____

15. How many roosters per 15 hens do you need to have fertile eggs?

_____

Underline the correct word

16. (Corn, Wheat) is necessary feed for good egg production.

17. Hatching eggs require special care (3 to 5 days, 7 to 10 days).

18. The chicken coop should be closed each night to (keep out, keep) dangerous animals.

19. Excess young roosters and other older birds that are no longer desirable for breeding should be (disqualified, eliminated).

20. The chicken coop of 5 meters is sufficient for (50, 100) adult chickens.

Fill in with the correct answer

1-3. What are the signs of Newcastle Disease?

  a. _____

  b. _____

  c.

4. The _____ is for the whole flock and needs to be repeated every _____ months.

5. The vaccine can be bought under the name _____ in a veterinary drug store.

6. When vaccinating, be careful not to place your hands near the face as the vaccine can infect _____ _____.

7. _____ any remaining vaccine with the container and _____ _____ thoroughly.

Fill in with the correct word

8. Chickens will be _____ and _____ when parasites are controlled.

9. Small _____ are the most seriously affected by internal _____ and some may even die.

10. It is best to _____ the entire flock.

11. Deworming pills can be purchased in veterinary drug stores under the name _____.

12. Lice should be controlled as they cause _____ and _____.

13. Lice can be controlled by mixing one part _____ powder with four parts of ashes.

14. All treatments should be done every _____.

Fill in the correct word

15. _____ is a common long lasting infection characterized by _____.

16. Sick birds lose their _____ and do not lay many eggs.

17. Chronic Respiratory Disease can be controlled with the antibiotic _____.

18. It is necessary to treat all sick and healthy birds, giving each a _____ in the mouth.

Review #2

Respond with a T for those answers that are TRUE and with an F for FALSE

1. ___ NEWCASTLE vaccine can be bought in any drug stores.

2. ___ Burn any remaining vaccine with the container and wash hands thoroughly.

3. ___ Chickens will be stronger and healthier when parasites are controlled.

4. ___ It is best to deworm one flock at a time.

5. ___ Lice should be controlled as they cause weight loss and higher egg production.

6. ___ All treatments should be done every three months.

7. ___ Chronic Respiratory Disease is a common long lasting infection characterized by nasal discharge.

8. ___ Sick birds has a lot of appetite and lays many eggs.

9. ___ Terramycin is an antibiotic that can control Chronic Respiratory Disease.

10. ___ Malathion powder controls lice.

Pre-Test

Underline the correct answer

1. Antibiotic a medicine given to treat against (disease, germs).

2. (Vaccination, Terramycin) a medicine given to an animal that will protect them from getting a disease.

3. (Contamination, Infectious) the effect of disease producing germs causing an infection.

4. (Everyone, A farmer) can raise backyard chickens.

5. Because eggs are rich in protein, it is good to serve an egg daily to children so that they will grow (strong and healthy, alert and active).

6. The best broody hen can cover (10 to 12, 13 to 15) eggs in the nest.

7. To have a successful natural incubation one should select well-formed, (medium to large, small to medium) sized eggs.

8. For fertile eggs there should be 1 rooster per (13, 15) hens.

9. The chickens should hatch (23, 21) days.

10. (Corn, Wheat) is necessary feed for good egg production.

Respond with a T for those answers that are TRUE and with an F for FALSE

11. ___ Give grilled corn to baby chicks which can eat whole grains.

12. ___ The chicken coop should remain open all day so that the hens can enter and lay eggs.

13. ___ The chicken coop should be closed each night to keep out dangerous animals

14. ___ It is necessary to maintain 2 roosters per 12 hens.

15. ___ It is a good practice to fill the boxes and nests with chopped sticks.

Fill in the correct answer

16-18 What are the signs of Newcastle Disease?

16. _____

17. _____

18. _____

19. The vaccine can be bought under the name _____ in a veterinary drug store.

20. Chronic Respiratory Disease can be controlled with the antibiotic _____.

21. Sick birds lose their _____ and do not lay many eggs.

22. Burn any remaining vaccine with the container and _____ thoroughly.

23. All treatments should be done every _____ months.

24. Lice should be controlled as they cause weight loss and _____.

25. When vaccinating, be careful not to place your hands near the face as the vaccine can infect _____ eyes.

Let's Review Key

IMPROVED BACKYARD CHICKEN PRODUCTION

1. F

2. C

3. A

4. G

5. B

6. D

7. J

8. E

9. H

10. I

11. Everyone

12. broody, hen, twelve

13. protein, strong, healthy

14. Chicken, eggs, family, income

15. Corn, 50

16. tasting, higher, hatchery

17. hatching, raising

18. scratching, insects, weeds

19. Local birds

20. Lifts its feathers

21. Makes a warning sound

22. Seldom leaves it nest

23. two or three

24. 10 to 12 eggs

25. one should select well-formed, medium to large sized eggs

26. 1 rooster per 15 hens

27. 10 to 12 eggs

28. mark the date on each egg with a pencil to avoid confusion with freshly laid eggs

29. 21 days

30. with sufficient ground corn and fresh water

31. ground

32. open

33. 7 to 10 days

34. keep out

35. eliminated

36. 15

37. predators

38. 50

39. cornhusks

Review #1

1. Vaccination

2. Incubation

3. Hatchery

4. Antibiotic

5. Predators

6. F

7. T

8. F

9. F

10. T

11. Lifts its feathers

12. Makes a warning sound

13. seldom leaves its nest

14. 10 to 12 eggs

15. 1 rooster

16. corn

17. 7 to 10 days

18. keep out

19. eliminated

20. 50

1. gasping

2. coughing

3. drooping wings

4. vaccination, three

5. Newcastle

6. human eyes

7. burn, wash hands

8. stronger, healthier

9. chickens, parasites

10. deworm

11. Triple wormer or wormal

12. weight loss, lower egg production

13. Malathion

14. three months

15. Chronic Respiratory Disease, nasal discharge

16. appetite

17. Terramycin

18. teaspoon

Review #2

1. F

2. T

3. T

4. F

5. F

6. T

7. T

8. F

9. T

10. T

Pre-Test

1. Disease

2. Vaccination

3. Infectious

4. Everyone

5. Strong and healthy

6. 10 to 12

7. Medium to large

8. 15

9. 21

10. Corn

1. F

2. T

3. T

4. F

5. F

6. gasping

7. coughing

8. drooping wings

9. NEWCASTLE

10. Terramycin

11. appetite

12. wash hands

13. three

14. lower egg production

15. human

Final Test

Improved Backyard Chicken Production Final Test

FINAL TEST

Respond with a T for those answers that are TRUE and with an F for FALSE

1. _____ Everyone can raise backyard chickens.

2. _____ Chicken and eggs are highly regarded foods and provides a way to improve income.

3. _____ A farm family can usually produce enough corn to feed up to 50 adult chickens.

4. _____ Local hens are excellent for hatching eggs and raising chicks.

5. _____ The broody hen can barely cover 5 to 6 eggs in the nest.

6. _____ For fertile eggs, there should be 2 roosters per 20 hens.

7. _____ Five hens can incubate 10 to 12 eggs.

8. _____ The chickens should hatch 18 days.

9. _____ Corn is necessary feed for good egg production.

10. _____ Give grilled corn to baby chicks.

Underline the correct answer

11. The chicken coop should be closed each night to (keep in, keep out) dangerous animals.

12. Hatching eggs require special care (3 to 5 days, 7 to 10 days).

13. The chicken coop should be strongly built so that night (owls, predators) can not enter.

14. Excess young roosters and other older birds that are no longer desirable for breeding should be (disqualified, eliminated).

15. A chicken coop of 5 meters is sufficient for (50, 100) adult chickens.

Fill in with the correct answer

16-18. What are the signs of Newcastle Disease?

_____

_____

Fill in with the correct word

Malathion	Lower Egg Production
Three	Terramycin
Appetite	Stronger
Healthier	Wash Hands

19. Burn any remaining vaccine with the container and _____ thoroughly.

20. Chickens will be _____ and _____ when parasites are controlled.

21. All treatments should be done every _____ months.

22. Lice should be controlled as they cause weight loss and _____ .

23. Lice can be controlled by mixing one part _____ powder with four parts ashes.

24. Sick birds loose their _____ and do not lay many eggs.

25. Chronic Respiratory Disease can be controlled with the antibiotic

_____ .

Final Test Key

IMPROVED BACKYARD CHICKEN PRODUCTION FINAL TEST KEY

1. T

2. T

3. T

4. T

5. F

6. F

7. F

8. F

9. T

10. F

11. Keep out

12. 7 to 10 days

13. Predators

14. Eliminated

15. 50

16. Gasping

17. Coughing

18. Drooping wings

19. Wash hands

20. Stronger, healthier

21. Three

22. Lower egg production

23. Malathion

24. Appetite

25. Terramycin

www.ingramcontent.com/pod-product-compliance
Lightning Source LLC
Chambersburg PA
CBHW081809200326
41597CB00023B/4203